Synthesis Lectures on Sustainable Development

Series Editor

Thomas Siller, Civil and Environmental Engineering, Colorado State University, Fort Collins, USA

This series publishes short books related to sustainable development practices relevant to engineers, technologists, managers, educators, and policy makers. The books are organized around the United Nations Sustainable Development Goals for 2015–2030. Design for sustainability along with the economics of sustainable development will be the common themes for each book. Topics to be covered will span all the major engineering disciplines along with the natural and environmental sciences that contribute to an understanding of sustainable development. The goal of this series is to make theory and research accessible to practitioners working on sustainable development efforts.

Hemantkumar N. Akolkar ·
Mubarak H. Shaikh · A. K. Haghi

Ionic Liquids

Greener Solvents for Sustainable Chemistry

Springer

Hemantkumar N. Akolkar
Department of Chemistry
Abasaheb Marathe Arts and New Commerce
Science College
Rajapur, Maharashtra, India

Mubarak H. Shaikh
Department of Chemistry
Radhabai Kale Mahila Mahavidyalaya
Ahmednagar, Maharashtra, India

A. K. Haghi
Edinburgh, UK

ISSN 2637-7675 ISSN 2637-7691 (electronic)
Synthesis Lectures on Sustainable Development
ISBN 978-3-032-02745-0 ISBN 978-3-032-02746-7 (eBook)
https://doi.org/10.1007/978-3-032-02746-7

This Springer imprint is published by the registered company Springer Nature Switzerland AG
The registered company address is: Gewerbestrasse 11, 6330 Cham, Switzerland

Preface

Ionic liquids have become a more environmentally friendly class of solvents in recent years, providing an alternative to conventional volatile organic compounds (VOCs). These ion-based liquids have special qualities that make them perfect for a variety of sustainable chemical applications. Ionic liquids have demonstrated great promise in lowering the environmental impact of chemical processes, from catalysis and synthesis to energy storage and environmental remediation.

With topics including the principles of ionic liquid chemistry, green synthesis techniques, and numerous applications, the book *Ionic Liquids—Greener Solvents for Sustainable Chemistry* seeks to give readers a thorough understanding of the discipline. Each of the book's four chapters focusses on a distinct facet of ionic liquids.

Chapter 1 introduces the basics of ionic liquid chemistry, including their structure, properties, classification, synthesis, and characterization. Chapter 2 explores green synthesis methods, highlighting the use of ionic liquids in catalytic reactions, organic synthesis, biocatalysis, and photochemical and electrochemical synthesis. Chapter 3 delves into the applications of ionic liquids, showcasing their potential in pharmaceutical and fine chemical synthesis, biomass conversion, energy storage and conversion, and environmental remediation. Finally, Chap. 4 discusses emerging trends and opportunities, including ionic liquid-based nanocatalysis, sustainable energy applications, biodegradable ionic liquids, and future directions in ionic liquid research.

Researchers, students, and professionals interested in sustainable chemistry and green technologies in the domains of chemistry, materials science, and engineering are the target audience for this book. We believe that anyone interested in learning more about how ionic liquids could contribute to a more sustainable future will find this book to be a useful resource.

We hope that readers will find this book informative, engaging, and inspiring, and that it will motivate them to explore the exciting world of ionic liquids.

Rajapur, India — Dr. Hemantkumar N. Akolkar
Ahmednagar, India — Dr. Mubarak H. Shaikh
Edinburgh, UK — Dr. A. K. Haghi

Contents

About the Authors

Hemantkumar N. Akolkar Ph.D., is currently working as an Associate Professor in the Department of Chemistry at Rayat Shikshan Sanstha's Abasaheb Marathe Arts and New Commerce Science College, Maharashtra, India. He is an academician and researcher who has worked for the last 14 years at both undergraduate and postgraduate levels. He has published five academic research-oriented books and 60+ research papers in national and international journals and conferences of repute. His areas of interest are heterocyclic chemistry, synthetic organic chemistry, and green chemistry. He is a reviewer of several journals of international repute in chemistry.

Dr. Mubarak H. Shaikh Ph.D., completed his doctoral study at Department of Chemistry, Dr. Babasaheb Ambedkar Marathwada University, Aurangabad, under the supervision of Prof. Bapurao Shingate in August 2016. His post-doctoral research work was carried out at Shandong University, Jinan, China where he worked with Prof. Wenguang Wang. In December 2017, he joined Wockhardt Research Center, Aurangabad as research scientist. In his two years of tenure at Wockhardt R&D, he successfully implemented improved process of Amidoxime, R-Nosylate and final API Nafithromycin core on plant scale up. Now, Dr. Mubarak H. Shaikh working as Assistant Professor at Department of Chemistry, Rayat Shikshan Sanstha's, Radhabai Kale Mahila Mahavidyalaya, Ahmednagar, Maharashtra, India. He is involved in conducting the research in the field of Organic synthesis with a focus on green catalytic synthesis and C-H functionalization, with an emphasis on the development of new synthetic methods that facilitate the construction of complex and bioactive molecules. After synthesis of the molecules, these molecules were tested for antimicrobial, antioxidant, anti-inflammatory, antitubercular, cytotoxic activity along with computational study. Till today, he had published more than 55 Research papers in the international repute along with eight Indian patents and eight book chapters included in RSC, Springer and Taylor & Francis publication. His research papers total citations are 1173, h-index is 18 and i10-index is 25. Having the perfect blend of Academics and Research; Dr. Mubarak has a long way to achieve many milestones in his life.

A. K. Haghi Ph.D., is a retired Professor of Engineering Sciences and has published 250+ academic research-oriented books and 1000+ research papers in various journals and conference proceedings. His leadership in academic publishing includes founding and serving as Editor-in-Chief of prestigious journals. Professor Haghi's extensive educational background and supervisory roles underscore his expertise and contributions to the field of engineering sciences. He is currently appointed as Honorary Research Associate (HRA) at University of Coimbra, Portugal.

List of Figures

List of Tables

1 Introduction to Ionic Liquids Chemistry

Abstract

Ionic liquids (ILs), which consist solely of ions, have turned out to be a special category of solvents with remarkable physicochemical characteristics and versatile utility in chemistry and materials science. Their structure and characteristics are thoroughly covered in this chapter, which also establishes the basis for their classification into a number of functional types, including task-specific, chiral, tunable polarity solvent, protic, basic, neutral, bio-ILs, poly(ILs), and supported ILs. Every class is made to perform particular functions in industrial, biological, and chemical processes. Modern characterization techniques like FTIR, XPS, NMR, TGA, and TEM are discussed, along with the synthesis of ILs. Viscosity, surface tension, ionic conductivity, solubility, polarity, thermal stability, and low vapor pressure are just a few of the precise physicochemical characteristics that make them ideal for use in green chemistry. In many fields of research and industry, ILs have evolved as solvents, catalysts, electrolytes, and functional materials due to their ability to control precisely these characteristics through cation and anion modifications.

1.1 Structure and Properties of ILs

ILs are a class of liquid salts that consist of cations and anions, typically exhibiting stability at or near ambient temperature. Their inherent tunability contributes to their versatility, allowing various applications across various systems (Fritsch and Strassner 2024). Walden et al. discovered the first IL in 1914, named ethylammonium nitrate, molten salt prepared at RT using ethylamine and concentrated nitric acid. However, due to its instability and explosive nature, it did not receive further attention at the time. Before his discovery, scientists everywhere believed ionic species required molecular solvents for

H. N. Akolkar et al., *Ionic Liquids*, Synthesis Lectures on Sustainable Development,
https://doi.org/10.1007/978-3-032-02746-7_1

liquid state existence because the established liquid theory involved uncharged molecules like water, benzene, and ethanol, along with atomic fluids such as mercury and argon. The unique property that made ethylammonium nitrate stand out was its high electrical conductivity because it contains mobile anions and cations. However, this compound also exhibited characteristics similar to water, such as a clear, colorless, and odorless appearance with a density resembling solids and fluid viscosity. Walden's previous research on electrolyte properties in water solutions revealed that ethylammonium nitrate primarily contains free ions, which established the basis for liquid salts. The field of IL science experienced significant advancement over the last three decades, while research in all branches of chemistry grew exponentially. Their development occurred through green chemistry principles (Hayes et al. 2015; Hussain et al. 2025).

ILs are now an alternative to organic solvents for different analytical methods, as they are more environmentally friendly than traditional organic solvents (David et al. 2022). Compared to conventional volatile organic solvents, they significantly improve laboratory safety by generating almost negligible vapor pressure. Due to their solvent properties and broad dissolution range for polar and non-polar compounds, they exhibit exceptional thermal stability, making them suitable for chromatography, spectroscopic, electrochemical investigations, liquid–liquid extraction, and solid-phase microextraction applications (Mastellone et al. 2023). They are considered green solvents as they have a non-volatile nature, thermal stability, and variable solvation power (Valli Sowbhagyam 2024). Compared to flammable and reactive volatile organic solvents, they are immiscible with air, so they do not cause air pollution; also, they do not evaporate to pose the risk of inhalation exposure. One of the advantages of using them for energy storage is that by adjusting the structure of the cation and anion, or both, it is possible to refine their thermal and electrical properties (Choudhary et al. 2023).

One of the main characteristics of ILs is their design capability, referred to as "tunability." To produce them with specific viscosity levels, polarity measurements, acidity qualities, conductivity abilities, and solubility properties, chemists alter the cation or anion structure or both. The design flexibility of solvents has introduced a novel method that makes it possible to synthesize solvent properties that perfectly align with the demands of reaction or separation systems (Aangothu et al. 2024). Modifying ILs according to specific requirements enables their broad utilization across different sectors, such as catalysis, electrochemistry, separation processes, and biomass processing. Their applications extend from pharmaceuticals to analytical chemistry and enzyme and protein stabilization (Mutelet 2023).

ILs consist of ions, and as they possess unique properties, they attract attention for various applications. Predicting their physical properties from ion structures is challenging (Tsuzuki and Shinoda 2023). Their structure differs from standard liquid arrangements because these liquids form structured internal patterns. Positively charged cations and negatively charged anions in these liquids develop powerful coulombic forces that attract each other strongly. Electrical forces make positive and negative ions join together to form domains called polar domains. Some cation and anion components in ILs possess longer parts that do not integrate properly with polar domains while being considered non-polar. The presence of lengthy non-polar tail structures leads to their arrangement into separate non-polar nanodomains, which are proximate to polar domains. Their properties, like flow rate, dissolving power, and electrical conduction, can be controlled or enhanced by studying how ion shapes and structures produce their internal structural complexity (Zhao et al. 2019).

Structurally, ILs consist of a cation and an anion. Generally, cations are large and organic, while anions are small and organic or inorganic. Asymmetry in their structure affects the physicochemical properties. Their ionic size asymmetry significantly impacts ionic density profiles and charge screening. According to the analysis study, asymmetric systems show higher coupling, influencing how they react to electrode polarization and encouraging spontaneous structures at interfaces (de Souza et al. 2022). Chemical structures of common cations are given in (Fig. 1.1).

Imidazolium, pyridinium, ammonium, or phosphonium derivatives are cations structurally based on nitrogen or phosphorus and commonly found in ILs. These cations have symmetry, and flexibility is essential in determining melting points; lower melting points result from more flexible structures since they have less lattice energy. Simple halides like Cl^- and Br^-, as well as complicated fluorinated species like bis(trifluoromethylsulfonyl)imide ($[NTf_2]^-$) and hexafluorophosphate ($[PF_6]^-$), can be anions. The low melting temperatures of ILs are due to the asymmetry and delocalization of charges in these ions, which inhibits the formation of solid crystalline lattices (Bernardino et al. 2020; Shukla et al. 2011).

The fundamental chemical components of ILs include organic imidazolium, pyridinium, or ammonium cations that link with multiple anions consisting of nitrate as well as halides and phosphate derivatives, and tetrafluoroborate (BF_4^-), trifluoromethanesulfonate, bistriflimide (Tf_2N^-), and hexafluorophosphate (PF_6^-). They comprise different cations and anions, which yield a wide range of physical characteristics by determining their solvent miscibility and positioning between hydrophobic and hydrophilic properties. The application requirements specify how the attached alkyl chain length controls hydrophobicity because it is a structural parameter for specific properties. Their flexible chemical structure enables their unique behavior to adapt across multiple chemical and industrial operations (Hassan et al. 2022).

The numerous possible combinations of ions that form ILs make it impossible to establish a single set of properties across all such compounds. Multiple physical and chemical characteristics emerge in each of them because of their cation–anion combinations. All of them share one essential trait, which is ionic conductivity brought about by mobile ions in their liquid state. They present solvent abilities that differ significantly from atomic, diatomic, and aprotic molecular liquids but share the common feature of being ionic (Nasirpour et al. 2020).

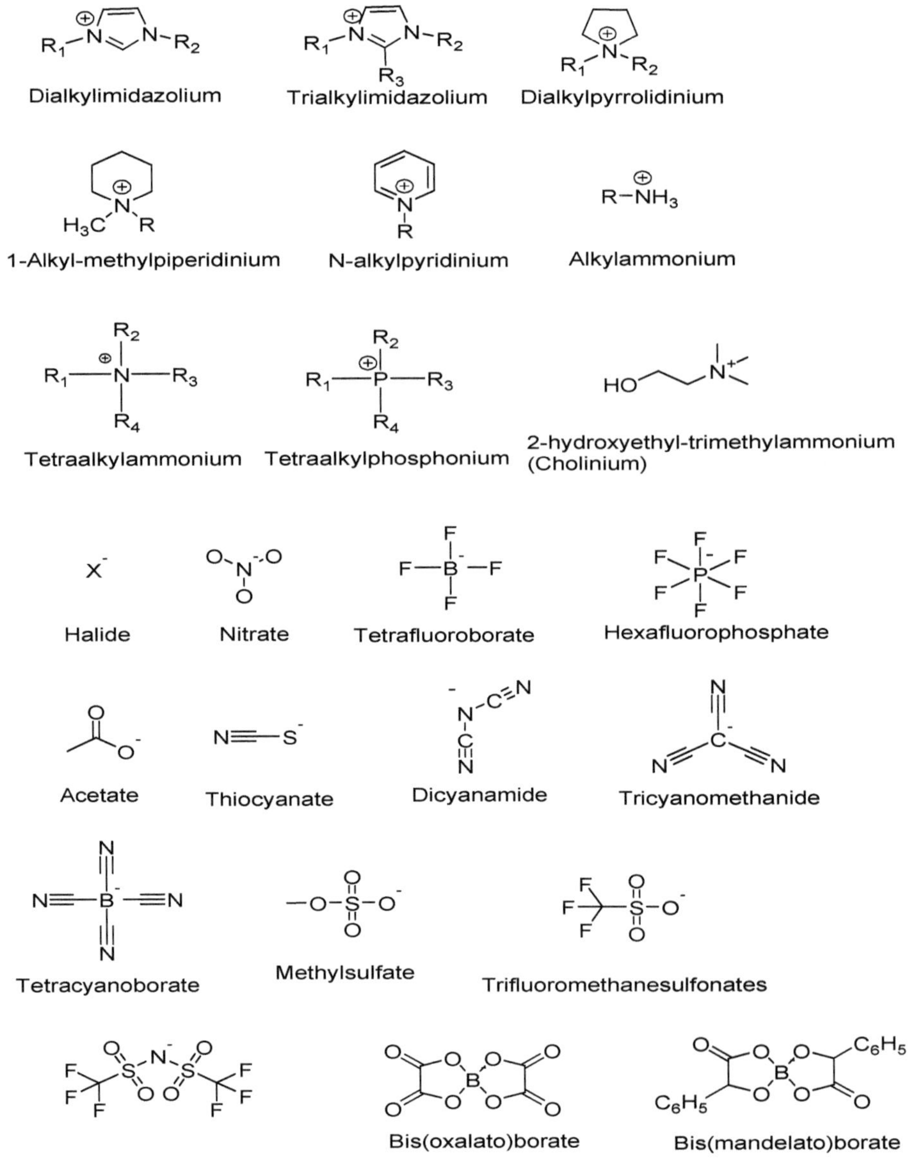

Fig. 1.1 Chemical structures of cations and anions

1.2 Classification of ILs

There are different types of ILs based on various parameters. Ammonium, phosphonium, imidazolium, triazolium, pyridinium, pyrazolium, and sulfonium are different cations in ILs, which are classified according to the presence of cations (A et al. 2023). Their properties heavily depend on the selection of anionic components. Although 2-hydroxy-*N*,*N*-dimethylethanaminium trifluoroacetate is an organic anion containing an IL compound, which shows reduced toxicity and improved biocompatibility, making it suitable for pharmaceutical and biomedical applications (Zhuo et al. 2024). The inorganic anions containing ILs [BMIM][$AlCl_4$] or [BMIM]Cl exhibit significant ionic character alongside high catalytic properties and excellent ionic conductance for applications including materials synthesis and catalysis (Shukla et al. 2023). On the availability of dissociable protons, they are protic and aprotic (Liu et al. 2024). Functionalized ILs are classified as hydroxyl-functionalized, carboxyl-functionalized, and amino-functionalized ILs (Qu et al. 2021; Ruan et al. 2025; Wang et al. 2023) (Fig. 1.2).

ILs classification depends on the variable cation–anion composition alongside their specific physical, chemical, thermal behaviors and biological properties and also on several key factors, including their water and organic solvent miscibility, their ability to conduct ions, and their acidic and basic properties, such as depicted in (Fig. 1.3). Varying ionic components generate flexible molecular structures, enabling these compounds to show distinct functional behaviors between highly conductive and hydrophilic to hydrophobic and chemically inert. Such structure variability permits their classification and targeted implementation across catalysis applications alongside electrochemistry, biotechnology, and materials science fields (Singh and Savoy 2020).

1.2.1 Task-Specific ILs

To accomplish specific reaction objectives, researchers alter the chemical composition by synthesizing a unique category of ILs. Task-specific ILs have typical structures composed of organic cations like imidazolium, pyridinium, or ammonium that form bonds with different anions and have functional groups covalently bonded. In order to build molecule affinities and achieve target-based chemical reactivity, functional groups are selected before anticipated molecular applications. They have unique properties that combine the typical advantages of ILs, such as low volatility, thermal stability, tunable polarity, and high ionic conductivity, with particular application functions like selective extraction and catalytic and molecular recognition activities (Llaver et al. 2022) (Table 1.1).

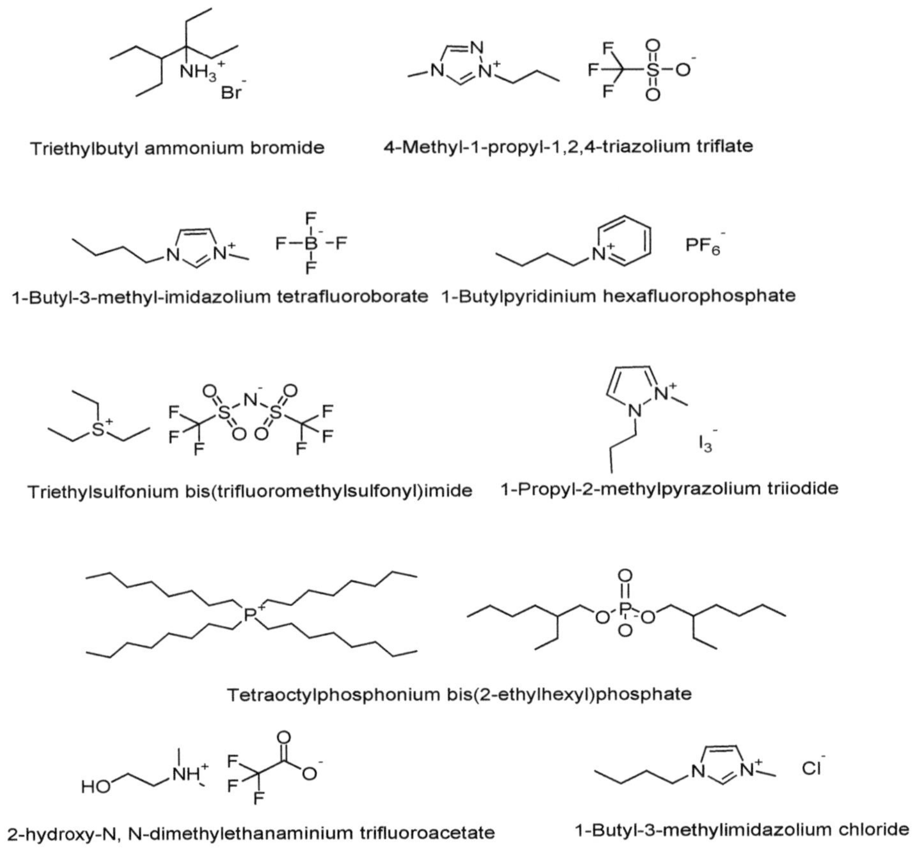

Fig. 1.2 Examples of different ILs based on cationic structures and organic/inorganic anions

1.2.2 Tunable Polarity Solvent ILs

Water and aqueous-based reagents were widely used as solvents, but ILs were attracted due to the limitations of conventional solvents (Bhowmik et al. 2024; Hatvate et al. 2018a, b, 2020; Patil et al. 2023). Switchable polarity ILs are a recent material innovation that solves existing problems in traditional SPS. Tunable solvents exhibit reversible changes in polarity in response to external stimuli, such as temperature or gas presence. Traditional SPS undergo evaporative losses; researchers have developed tunable polarity ILs to accomplish low-volatility polarity switching along with the same beneficial reversible polarity features. The structural design of these systems allows for the uniting of polarity adjustment capabilities of SPS with the volatility reduction, chemical and thermal resistance of typical ILs (Chakraborty et al. 2022; Jain et al. 2022). The combination of

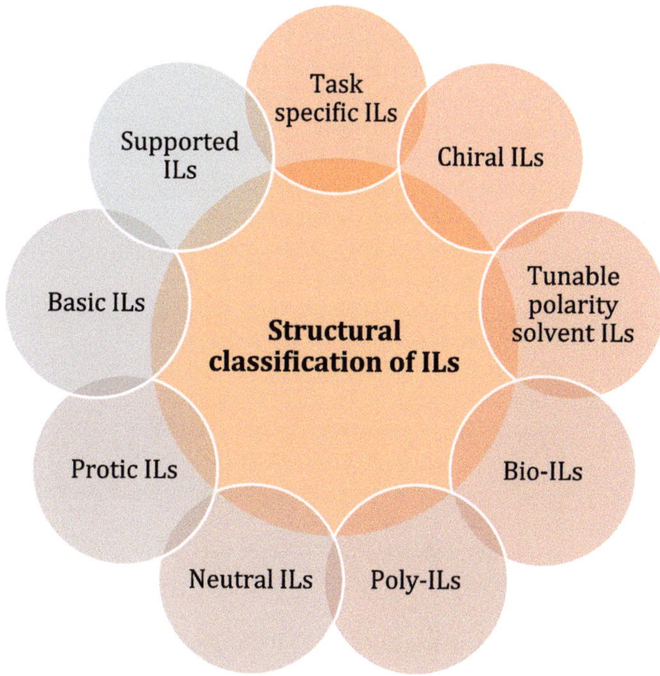

Fig. 1.3 Classification of ILs

1,8-diazabicyclo[5.4.0]undec-7-ene (DBU) and alcohol results in a low-polarity mixture, whereas its reaction with CO_2 produces a high-polarity IL (Singh and Savoy 2020).

1.2.3 Chiral ILs

Chiral ILs represent a specific subclass of ILs as they contain chirality within their anionic and cationic structure or across both components. They obtain chiral recognition capability from their inherent chirality, which is required to perform enantioselective separation and synthesis procedures. They also retain all typical qualities of conventional ILs while offering specific characteristics that provide exceptional enantiomer-separating power, increased polarity levels and solubility, and can be recycled safely for the environment (Dong et al. 2023).

Table 1.1 Classification of task-specific ILs

Types of task-specific ILs	Description	Examples	References
Amino-functionalized ILs	They contain an amino (NH_2) functional group in a cation/anion structure. Generally, the amino group is from amino acids	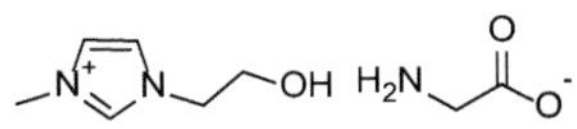 1-Hydroxyethyl-3-methylimidazolium glycinate	Lu et al. (2025)
Hydroxyl-functionalized ILs	Contains a hydroxyl (-OH) functional group with better H-bonding and solubility	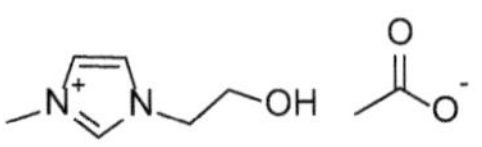 1-(2-Hydroxyethyl)-3-methylimidazolium acetate	Zhao et al. (2023)
Carboxyl-functionalized ILs	Carboxylic acid (−COOH) is a functional group present	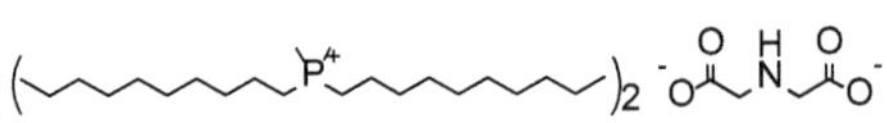 Didecyldimethylphosphonium iminodiacetate	Al-Sheakh et al. (2022)

1.2.4 Bio-ILs

Bio-ILs represent an environmentally friendly class that originates from combining bio-based positive and negative ions or both components. They involved the combination of cholinium cation, which exhibits low toxicity and high biodegradability, with plant-derived carboxylic acids as anionic components. They maintain the regular essential features, which include low volatility, high thermal stability, and strong solvent properties. They also provide additional sustainability benefits and adverse effect reduction. Choline is a common cation with different anions like abeitrate, ferulate, caffeate, and coumarin-3-carboxylate (Sharma et al. 2021) (Fig. 1.4).

1.2.5 Poly(ILs)

A type of ion-containing polymer known as poly (ILs) is distinguished by adding IL moieties to its repeating units. Due to their structural integration, they have special physicochemical characteristics that set them apart from other ion-containing polymers like poly (sodium styrene sulfonate). When IL structures are incorporated into the backbone

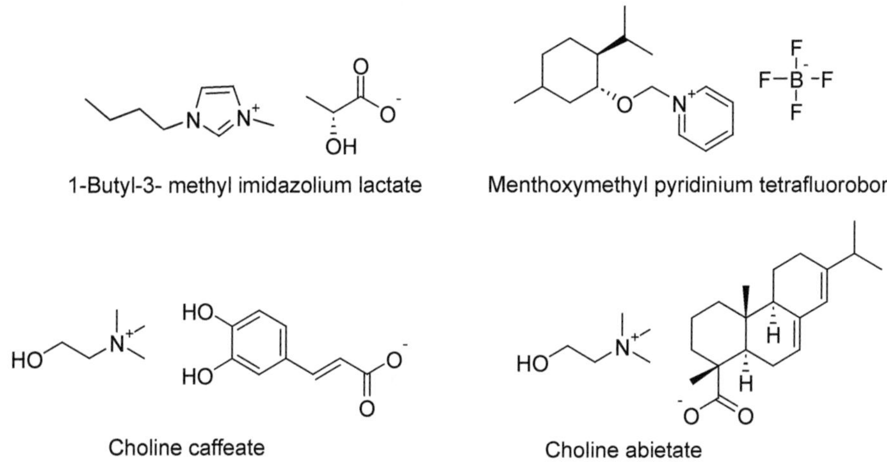

Fig. 1.4 Examples of chiral and Bio-ILs

of polymers, the materials have characteristics of both polymers, such as mechanical resilience and processability, as well as strong ionic conductivity and thermal stability (Matsumoto 2024).

1.2.6 Protic ILs

A transferable proton, usually produced by protonating a Bronsted base with a Bronsted acid, is a characteristic of protic ILs. A cation–anion pair that can form an H-bond due to this proton transfer significantly impacts their physicochemical characteristics. These H-bonding interactions provide distinct behaviors that affect their viscosity, conductivity, and thermal stability in contrast to aprotic ILs. They are utilized as electrolytes with strong ionic conductivity and thermal stability in energy storage devices like batteries and supercapacitors because of these properties (Stettner and Balducci 2021).

1.2.7 Basic ILs

Basic IL is a class that incorporates basic functional groups into its structure, enabling it to act as both solvents and catalysts in base-catalyzed reactions. Basicity can be linked to the cation or the anion while preparing basic ILs. However, the instability of some ionic species under basic conditions has made it challenging to prepare them. Their stability has been improved recently by material-based techniques such as heterogenization on solid supports and molecular design (McNeice et al. 2021).

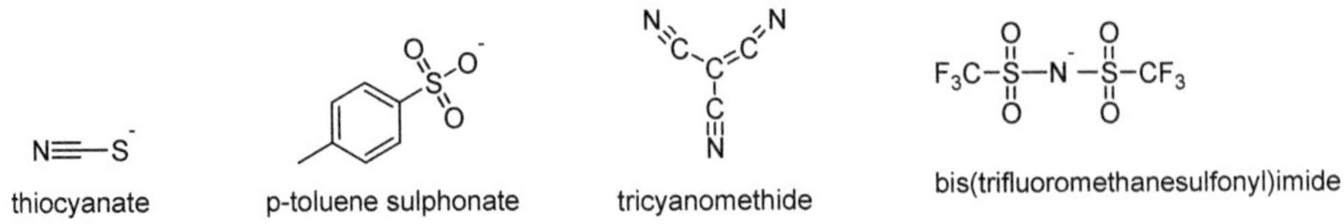

Fig. 1.5 Counteranions in neutral ILs

1.2.8 Neutral ILs

Counteranions of neutral ILs are developed with extremely weak electrostatic interactions to counteract cations, resulting in extremely low viscosity and melting temperatures. They are applied as inert solvents. Hexafluorophosphate, *p*-toluene sulphonate, tricyanomethide, thiocyanate, and bis(trifluoromethanesulfonyl)imide are some counteranions in neutral ILs (Singh and Savoy 2020) (Fig. 1.5).

1.2.9 Supported ILs

ILs immobilized on solid supports, such as silica, magnetic nanoparticles, or carbon nanotubes, are known as supported ILs. This immobilization leads through a physical adsorption, forming a supported IL phase, or through covalent bonding, forming a supported IL-like phase. Due to their affordability, surface hydroxyl groups that aid in functionalization, and clearly defined pore structures, silica-based supports are especially preferred. These characteristics make them perfect matrices for supported ILs in various catalytic applications (Lakhani et al. 2024).

1.3 Synthesis and Characterization of ILs

1.3.1 Synthesis of ILs

The synthesis of ILs involves two steps;

(a) Cation formation: Cation synthesis is done in two ways, either by protonating an amine using an acid or by practicing quaternization reactions with an alkyl halide and providing heat treatment.
(b) Exchange of anion: Anion exchange processes may be performed by using anion metathesis or by utilizing salts of halide with Lewis acids to form Lewis acid-based IoLs. Anion exchange is done by either a metal salt, a Bronsted acid, or a resin (Ratti 2014) (Scheme 1.1).

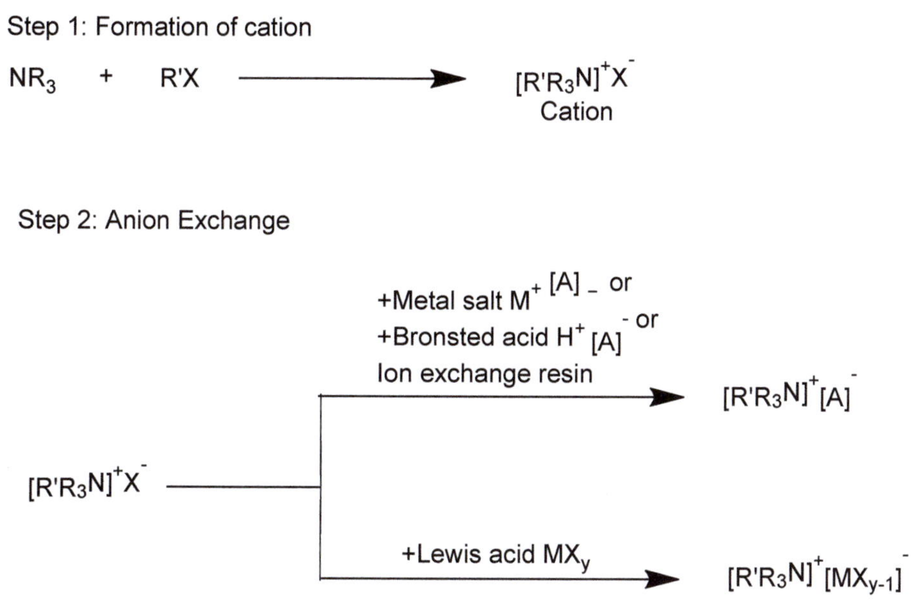

Scheme 1.1 Synthesis of ILs

A suitable method for preparing task-specific ILs is quaternizing a nucleophile (N-heterocycle, phosphine, etc.) and a functionalized haloalkane. The above displacement reaction is then followed by anion exchange. All IoLs that are stable towards bases can be synthesized by this method (Chiappe and Pomelli 2014) (Scheme 1.2).

Imidazolium-based ILs are prepared through two steps: cation formation and anion formation. Cation is formed from *N*-methylimidazole and an alkyl halide; further, anion exchange is carried out (Noorhisham et al. 2021) (Scheme 1.3).

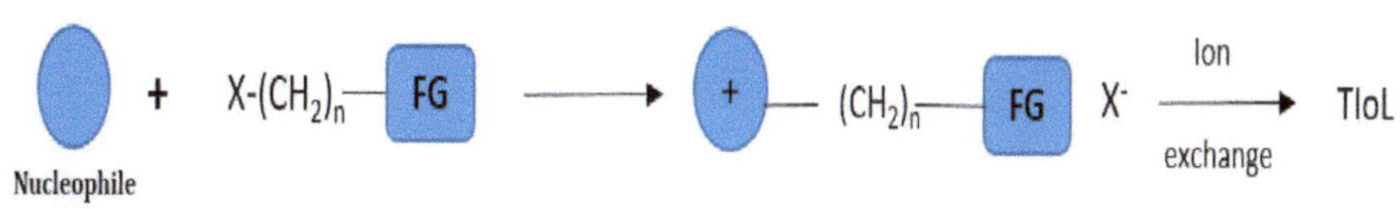

Scheme 1.2 Synthesis of task-specific ILs

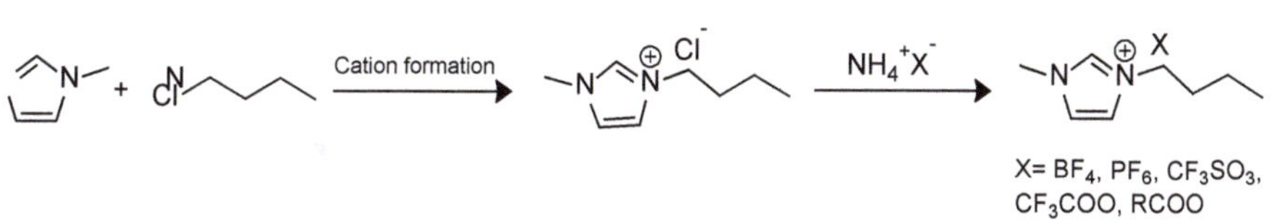

Scheme 1.3 Imidazolium-based ILs

A pyridinium-type IL is synthesized through two steps, i.e., cation formation and anion exchange. In the first step, *N*,*N*-dimethyl pyridine-4-amine was treated with benzyl bromide in the presence of ACN, and in the second step, anion exchange was carried out in the presence of counter anions (Tamilarasan et al. 2023) (Scheme 1.4).

The acid-base neutralization reaction prepares ammonium-based ILs. This process starts with adding an appropriate acid to a selected solvent to develop thiazolium-based ILs. After the acid has completely dissolved and been distributed evenly, an ammonium base is gradually added to the reaction mixture. Depending on the desired structure, this slow addition ensures complete neutralization or quaternization while reducing the possibility of adverse reactions and enabling regulated reaction kinetics (Tanzin et al. 2023) (Scheme 1.5).

In synthesizing thiazolium-containing ILs, an alkylating agent combines with the nitrogen atom of the thiazole ring to generate a thiazolium salt in the first step, known as quaternization, of a thiazole derivative. A stable cationic species is produced when this reaction gives the nitrogen a positive charge. An anion exchange is carried out in the second stage to modify the physicochemical characteristics of the IL by substituting a desired anion, such as bis(trifluoromethanesulfonyl)imide, for the original counterion, which is frequently a halide (Fei et al. 2021) (Scheme 1.6).

1-Hexylpyridin-1-ium 4-hydroxybenzoate is synthesized by quaternization and anion exchange. Quaternization of pyridine is carried out with hexyl bromide to form 1-hexylpyridin-1-ium bromide. Further anion exchange is done using naturally occurring

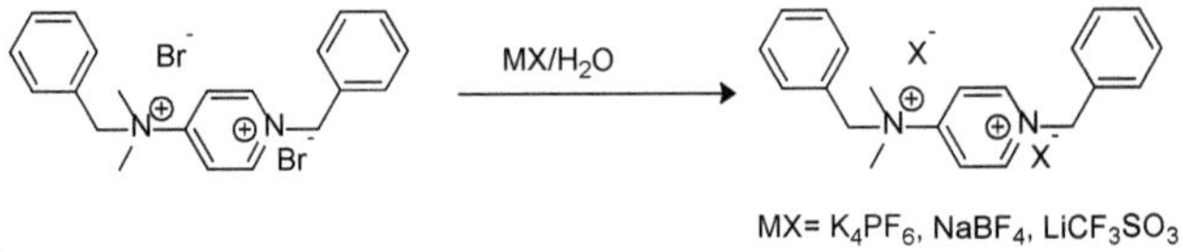

Scheme 1.4 Synthetic scheme for pyridinium-based IL

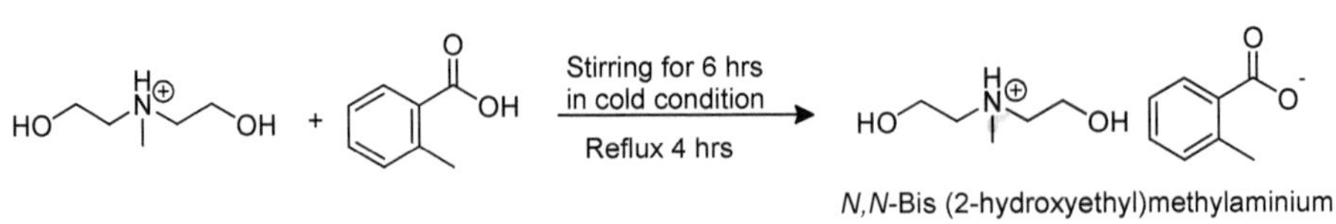

Scheme 1.5 Ammonium-based IL synthetic scheme

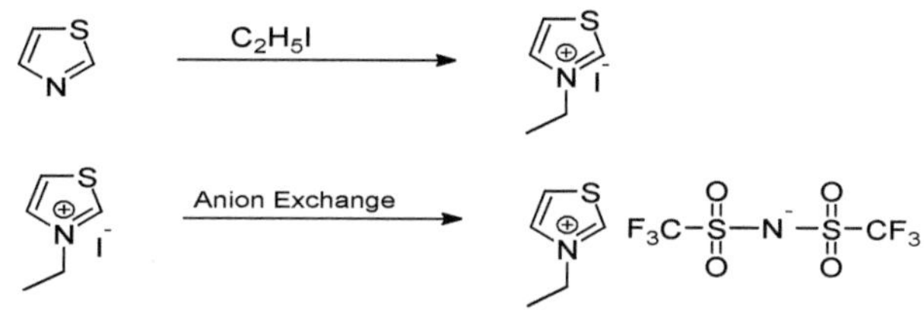

Scheme 1.6 Synthetic scheme for thiazolium-based ILs

benzoic acids with $NaHCO_3$, which is optimized through sodium salts with bromide (Javed et al. 2025) (Scheme 1.7).

For the synthesis of poly(ionic) liquids, poly(4-vinylpyridine) can be utilized. To achieve the poly(ionic) liquids monomer synthesis, polymerization, addition of Bronsted acid, and addition of Lewis acid are the four subsequent steps (Guzmán-Pantoja et al. 2025) (Scheme 1.8).

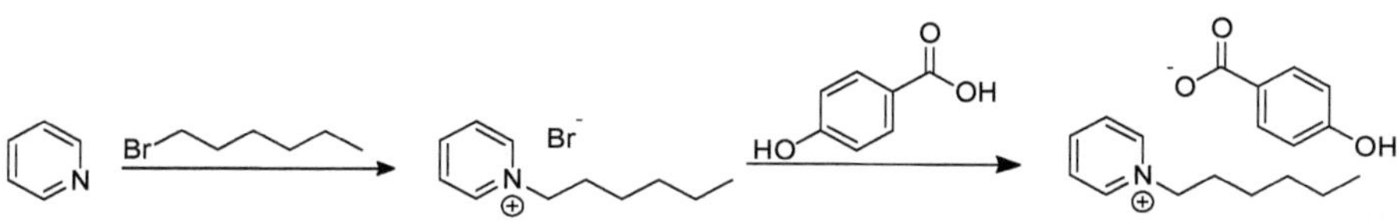

1-Hexylpyridin-1-ium 4-hydroxybenzoate

Scheme 1.7 Synthesis of 1-hexylpyridin-1-ium 4-hydroxybenzoate

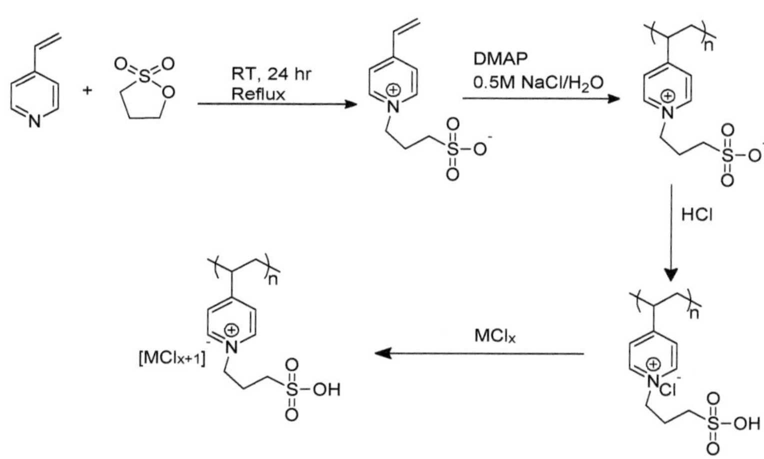

Scheme 1.8 Synthetic scheme for poly(ionic) liquids

1.3.2 Characterization of ILs

ILs must be characterized in order to better understand their thermal, chemical, and physical characteristics, all of which have a direct impact on how efficiently they function in different applications. A variety of analytical methods are needed to investigate their structure, stability, and dynamics because of their distinct ionic nature, low volatility, and potent intermolecular interactions. These methods offer important insights into temperature decomposition, phase behavior, molecular interactions, and nanoscale organization. The design of task-specific ILs with the required characteristics is also aided by accurate characterization. Researchers can obtain an extensive understanding of how ILs behave under various circumstances by combining several analytical approaches.

1.3.2.1 FTIR

Functional groups in ILs can be identified using FTIR spectroscopy by examining distinctive vibrational modes. Studying hydrogen bonds, particularly with regard to water content, is made easier with the help of FTIR. Inbaraj et al. demonstrate that the water activity is directly correlated with H-bond basicity. This discovery helps to clarify the hydrophobicity and their function in stabilizing biomolecules (Rajapriya Inbaraj et al. 2023). Temperature-dependent FTIR spectra are monitored for H-bonding, which is crucial to know the phase behavior and structural dynamics, especially protic ILs such as ethylammonium nitrate. Systematic shifts and variations in the width and intensity of distinctive vibrational bands, particularly in the N–H and O–H stretching areas, were noted as the temperature increased. The changes observed in spectra indicated both weakening of H-bonds and ionic network realignment, which produced alterations that revealed structural rearrangement ahead of phase change transitions. IR spectral features function as sensitive indicators that differentiate between solid, supercooled, and liquid phases (Palumbo et al. 2020) (Table 1.2).

Table 1.2 Common functional groups with their IR ranges

Functional groups	Wavenumber (cm^{-1})
−OH	3200–3600
C–H (Alkyl chains)	2850–2960
C=O (Carbonyl)	1650–1750
S=O (Sulfonyl)	1030–1350
P=O (Phosphoryl)	1200–1300
C=C (Aromatic)	1450–1600
C=N	1550–1650
N–H	3200–3500
C–F, B–F, or S–F (in Anions)	1000–1400

1.3.2.2 XPS

The XPS system functions as a crucial surface-sensitive analytical method for analyzing material composition and electronic states, together with their chemical bonding conditions. ILs demonstrate excellent suitability for XPS analysis because of their nearly nonexistent vapor pressure and their ionic behavior, combined with their low volatility properties. XPS provides comprehensive information about the cationic and anionic components by identifying changes in binding energy that correlate to modifications in the electronic surroundings. This is especially helpful for examining interionic interactions such as charge delocalization, ion pairing, and hydrogen bonding. Furthermore, under operational conditions such as during catalysis or electrochemical reactions, XPS can follow surface fragmentation events, distinguish between functionalized and non-functionalized ILs, and track variations. XPS is a vital tool for improving the comprehension and design of task-specific ILs for a variety of applications because of its capacity to resolve such surface-level information (Greco et al. 2019; Morales-Ugarte et al. 2019).

Gousseva et al. utilized XPS for the study of 1-butyl-3-methylimidazolium thiocyanate, a joint experimental and computational technique was used for analysis of complex electronic relations in this IL. The XPS analysis revealed specific binding energies of every important atomic site from both the imidazolium cation, along with S and N atoms in the thiocyanate anion, and from carbon and nitrogen atoms within the imidazolium cation (Gousseva et al. 2022).

1.3.2.3 NMR

Solution NMR techniques and solid-state NMR determine multiple aspects within ILs by combining their analysis methods. NMR experiments that examine different nuclei, including ^{1}H, ^{13}C, ^{19}F, ^{31}P, and ^{11}B, allow researchers to study individual ion surroundings, especially when investigating ILs with these heteroatoms. Different NMR spectroscopy techniques, such as NOESY and ROESY, reveal the spatial positions and chemical interactions between ions, thus helping researchers to understand the microstructures of ILs (Zhu and O'Dell 2021).

The practice of NMR relaxometry requires measurements between spin-lattice (T_1) and spin–spin (T_2) relaxation times to detect molecular dynamics. T_1 and T_2 relaxation methods in ILs enable the collection of data about the mobility of ions as well as their rotational motions and intermolecular forces between cations and anions. The combination of ^{1}H and ^{19}F relaxation methods allows both the determination of translational diffusion coefficients and rotational correlation times to study their dynamic behavior across different time domains (Endo et al. 2020). The self-diffusion coefficients determined by DOSY NMR help evaluate the sizes of molecules and interactions between molecules present in the solution. This technique helps to understand the distributions of individual ions alongside ion pairs and large molecular aggregates through diffusion measurement. By using DOSY NMR, imidazolium-based ILs undergo dynamic-structural behavior studies, resulting in observations about ion pairing and aggregation across various solvents (Zanatta et al. 2019). NOESY investigations reveal spatial proximities and molecular conformations by detecting through-space interactions between nuclei. In order to get a better understanding of microstructures and rotational dynamics, NOESY has been used to investigate inter-ionic interactions and the spatial organization of ions in ILs (Damodaran 2022).

Khudozhitkov et al. reported an isotope effect for triethylammonium methanesulfonate with the help of deuterium NMR. When it comes to dynamics in deuterated ILs, deuteron ^{2}H NMR is especially helpful. Researchers can examine dynamic heterogeneities and phase behaviors by examining ^{2}H spectra, which show anisotropic and isotropic signals corresponding to various mobility regimes. These investigations have shed light on the mobility and rotational dynamics of certain ions inside ILs (Khudozhitkov et al. 2022).

1.3.2.4 Thermogravimetric Analysis

One essential method for evaluating the thermal stability and the decomposition behavior of ILs is thermogravimetric analysis. Standard TGA measures the sample weight loss at various heating stages to determine stability patterns and decomposition points. The thermal analysis technique TGA detects moisture levels and decomposition activation points in ILs. Under a nitrogen atmosphere, the thermal stability measurement of 1-butyl-3-methylimidazolium bis(trifluoromethylsulfonyl)imide indicated decomposition starts at 672.65 K (Ravula et al. 2019). The combination of TGA with mass spectrometry enables decomposition product identification. The generation of free chloride during decomposition was indicated by the TGA-MS analysis of $[C_8C_1Im]Cl_xZnCl_2$ ILs, which showed the release of methyl chloride on heating (Clarke et al. 2022).

1.3.2.5 TEM

ILs can be studied using different TEM imaging modes. Significant improvements in their study have been made possible using TEM. The imaging of nanostructured domains, which reveals heterogeneities that affect characteristics like viscosity, conductivity, and solvation, is possible. Because of the amphiphilic nature of their pairs of cation-anion, ILs frequently exhibit microphase separation, which is readily observable using high-resolution TEM. Additionally, TEM has been crucial in capturing the nucleation and growth processes in real time, and they have been used as a stabilizing medium for the synthesis of metal nanoparticles. This advances our knowledge of how ILs stabilize nanoparticles and the processes underlying their self-assembly (Meischein and Ludwig 2021; Tie et al. 2023).

1.4 Physicochemical Properties of ILs

The distinct physicochemical characteristics of ILs are attributed to their composition of diverse anions and large organic cations. These include a broad range of liquid temperatures, great thermal stability, low vapor pressure, and strong solubility for a variety of inorganic and organic substances. Because of their adjustable structure, which enables customization of characteristics like density, polarity, and viscosity, they are beneficial in a variety of chemical and industrial applications. Their physicochemical properties depend on the nature of cations and anions present in the structure.

SAR of ILs

(a) **Nature of cation**

Asymmetric cations reduce lattice energy and lead to a lowering of the melting point of ILs. The melting point typically drops as the alkyl chain length of the cation grows from methyl to octyl, and it begins to rise once more with longer chains (such as C_{14} or above). Branched alkyl chains indicate less thermal stability, and long chains decrease their density, surface tension, and water solubility. Alkyl chain length tends to increase the rate of biodegradation when it contains more oxidizable carbons (Sharma et al. 2024). Compared to their alkyl-substituted counterparts, ILs with ether groups added to their side chains usually have higher fluidity and lower viscosity (Philippi et al. 2020). Functionalization of cation with hydroxyl or carboxy functional groups results in the formation of H-bonds (Hunger et al. 2022).

(b) **Nature of anion**

Halides containing ILs possess low thermal stability. Acetate, nitrate, and halides are hydrophilic anions that increase the water solubility, whereas $[PF_6]^-$ and $[NTf_2]^-$ are hydrophobic anions that impart water insolubility (Sharma et al. 2024). Fluorinated organic anions increase thermal stability and decrease the surface tension (Prikhod'ko et al. 2020). Basicity of anion decreases thermal stability and increases CO_2 absorption (McNeice et al. 2021). There is a good distribution of negative charge in the trihalide anion, so it possesses high ionic conductivity and low viscosity (Abu Talip et al. 2022).

(c) **Cation-Anion interaction**

The melting point of ILs is affected by the cation-anion interaction. Compared to alkyl-methylpyrazolium iodide, alkyl-methylpyrazolium triiodide ILs have substantially lower melting points due to the degree of the cation–anion interaction. The iodide anion has a

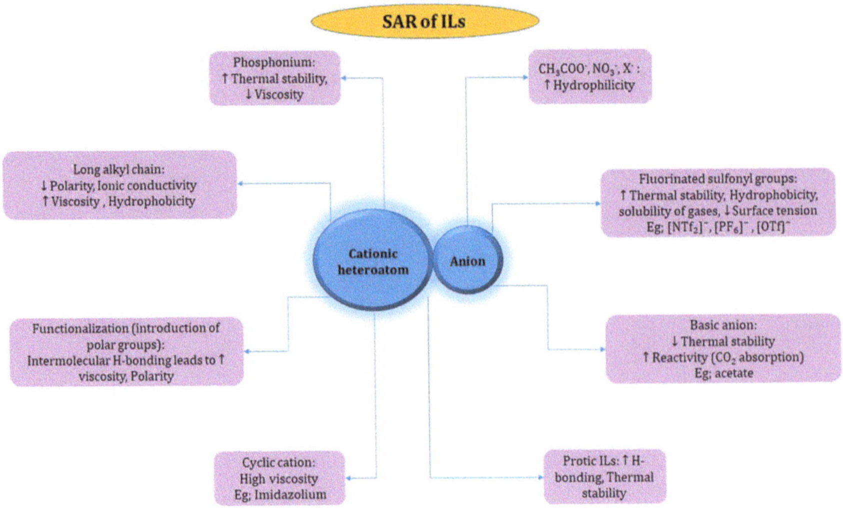

Fig. 1.6 SAR of ILs

more localized charge than the triiodide anion, which attracts a positively charged pyrazolium cation (Abu Talip et al. 2022). Hydrogen bonding, van der Waals interactions, polarization effects, and electrostatic attraction are some of the forces that interact between cations and anions in ionic liquids. Formerly, strong electrostatic interactions within ion pairs have been largely responsible for the intrinsically high viscosity of ILs (Ma et al. 2021) (Fig. 1.6).

1.4.1 Viscosity

The most important property that defines ILs is their viscosity. Higher viscosities could be preferred as a membrane phase or for lubrication; however, lower viscosities are generally preferred when they are used as solvents to minimize pumping costs and maximize mass transfer rates. This fact has already been cited earlier, stating that the kinds of cations and anions greatly affect their viscosity. The viscosities vary, and when measured at RT, they may vary between 10 and over 105 cP (Gardas and Coutinho 2008).

1.4.2 Surface Tension

The molecular structure of ILs, including the type of cation, anion, and any functional groups present, is intrinsically related to surface tension. They usually have higher surface tensions than traditional molecular solvents because of their ionic composition and potent

Table 1.3 Properties of ILs

Name of properties	Ranges	References
Density (kg/m^3)	8×10^{-4}–3.3×10^{-3}	Nasirpour et al. (2020)
Viscosity (Pa s)	0.014–40	
Refractive index	1.5–2.2	
Surface tension	18.5–70.3 mN/m	Mousavi et al. (2021)

intermolecular interactions. Their behavior in applications like electrochemistry, separation procedures, and catalysis is influenced by this characteristic (Mousavi et al. 2021) (Table 1.3).

1.4.3 Ionic Conductivity

As ILs are used as electrolytes in various electrochemical processes, it is important to know their ionic conductivity. The pure, non-clustered structure allows for free ion movement and good charge transport, which is necessary for energy storage applications (Labeeb et al. 2024). They have high ionic conductivity because of their inherent characteristic as salts having melting temperatures lower than 100 °C. This conductivity is due to the mobility of ions being considerably high due to the fluidity at room or slightly higher temperatures; their structure consists of large asymmetrical ions that reduce intensive ionic relations and do not form a strictly ordered crystal lattice, which is why the ions are freely movable. The system is inherently ionic, and due to the presence of ions, one is assured of high charge carriers (Ma et al. 2023). The structural modifications performed through changes to their cations or anions influence their ionic conductivity by modifying their viscosity along with ion mobility characteristics. The anionic variation between imidazolium-based ILs influences their conductivities, whereby [EMIM][BF_4] demonstrates higher conductivity levels than [BMIM][PF_6] because of anionic interaction differences and viscosity variations (Ranjan et al. 2021).

1.4.4 Solubility

One essential physicochemical characteristic of ILs that greatly affects their efficacy as solvents, particularly in medicinal applications, is solubility. The bioavailability and therapeutic efficacy of many pharmacological compounds are limited by their poor solubility in water and traditional organic solvents. Due to their adaptable structures, which enable fine-tuning solvation capabilities, they have become promising substitutes. Hydrophilic and coordinating anions like acetate and methylphosphonate have strong H-bonding interactions that improve drug dissolution. The solubility of a medication is mostly determined by the type of anion in the IL. On the other hand, those that contain hydrophobic and non-coordinating anions, such as bis(trifluoromethylsulfonyl)imide ($[NTf_2]^-$) or tetrafluoroborate ($[BF_4]^-$), frequently exhibit low solubilizing power (Adawiyah et al. 2016).

1.4.5 Polarity

Modifications in cation structure significantly impact the polarity of ILs. Alkyl substitution and the addition of hydroxyl (−OH) groups to the cation can impact the overall polarity. In particular, it has been demonstrated that adding −OH groups to 1-ethyl-3-methylimidazolium-based ILs greatly increases their polarity and reliance on the nature of anions. On the other hand, polarity usually decreases when the length of the alkyl chain on the cation or the anion is increased. The polarity is only moderately affected by variations in alkyl chain length, but it increases significantly when −OH groups are substituted (Yalcin et al. 2020).

1.4.6 Tunability

The exceptional chemical structural tunability of ILs enables the customization of their characteristics for particular biomedical uses. The capacity to modify their hydrophilicity or lipophilicity to improve penetration through biological membranes is one of their primary characteristics. When it comes to delivering APIs, surface-active ILs have shown better performance than traditional surfactants, making them promising drug carriers. Researchers have also demonstrated that they can alter the fluidity of membranes. Those based on hydrophobic imidazolium have a tendency to disrupt lipid bilayers and form temporary channels, which makes it easier for APIs to pass through membranes. On the other hand, hydrophilic ILs typically prevent such leakage and stabilize membrane structures (Shukla et al. 2023).

1.4.7 Low Melting Point

ILs are liquid at RT, remain fusible over a relatively wide range, and can be used at extremely low or high temperatures (neamah et al. 2022). Compared to “normal salts”, they show relatively low melting points. One of the most important aspects is the difference between the enthalpy and entropy of fusion. The extensive and unsymmetrical nature of the ions lowers the lattice energy and makes the packing more random, making the enthalpic contribution of the solid-state less effective than that of the liquids. Additionally, due to the flexibility of alkyl chains and the interactions between the anions and cations, molecular dynamics studies of specific ILs play a significant role in rotational/vibrational entropy, contributing to melting. Configurational entropy is an essential factor that contributes to low melting points by favoring a disordered state (Endo et al. 2022).

1.4.8 Thermal Stability

ILs are more thermally stable compared to organic solvents and can be used at considerably higher temperatures, over 400 °C. The anion has a greater liability for controlling the thermal stability than the cation. Here, the thermal stability depends on the hydrophilicity of the anion (neamah et al. 2022). ILs with fluorinated anions, such as bis(trifluoromethylsulfonyl)imide ($[NTf_2]^-$), have remarkable thermal stability, with breakdown temperatures exceeding 350 °C. This is due to charge delocalization and the inert nature of the sulphonyl groups. In contrast, those containing halide anions, such as chloride, break down at substantially lower temperatures due to their tendency to participate in side reactions or hydrolysis (Paolone et al. 2022; Prikhod'ko et al. 2020). $[BMIM][NTf_2]$ exhibited thermal stability over 400 °C, while [BMIM]Cl deteriorated below 200 °C under the same conditions. The structural characteristics and chemical stability of the anion are, therefore, essential factors that determine their thermal profile (Brunetti et al. 2023). The most thermally stable ILs contain cations such as imidazolium and pyrrolidinium, whereas the least thermally stable contain ammonium and pyridinium. Imidazolium-based ILs have a thermal stability range of 499 K (for $[C_2mim][AcO]$) to 732 K (for $[C_4mim][NTf_2]$). Ammonium-based ILs have a thermal stability of 423 K (for $[N_{4,4,4,4}][IO_3]$) to 505 K (for $[N_{4,4,4,4}][SCN]$) (Fabre and Murshed 2021).

1.4.9 Low Vapor Pressure

An aspect that has particularly received attention is a very low vapor pressure attributed to the ionic nature of ILs, and the coulombic forces between cations and anions limit their volatility (Pei et al. 2022). Their designation as non-volatile solvents is supported by recent thermogravimetric investigations, which demonstrate that the very slight volatility endures up to high temperatures. The fact that they do not release VOCs into the atmosphere is a significant factor in their classification as green solvents. Therefore, low vapor pressure improves safety and environmental compatibility in industrial and laboratory applications (Valli Sowbhagyam 2024).

1.5 Conclusion

ILs are tunable, green solvents with diverse applications in green chemistry, catalysis, electrochemistry, and material science. Here, their structure, types, methods for synthesis and characterization with physicochemical properties are discussed. Their properties and the extent of applicability rely on their structural design and functionalization. Furthermore, modern synthesis methodologies and characterization techniques provide the necessary tools to study and optimize ILs for specific industrial, analytical, and biological applications. In general, ILs present an exciting basis of invention in various fields of science. Overall, the chapter provides a substantial ground to perceive them as versatile and sustainable materials in contemporary science and industry.

References

A, S. M. S. H., A, A. A. A., A, O. S. A., A, M. M., A, A. M., B, H. J. K., B, M. M., & A, M. S. K. (2023). A review of ionic liquids: Recent synthetic advances and oilfield applications. *Journal of the Taiwan Institute of Chemical Engineers, Volume 153.* https://doi.org/10.1016/j.jtice.2023.105195

Aangothu, S. R., Prathipati, S., Chintala, J. N. P. K., C., S., & Bollikolla, H. B. (2024). A Comprehensive Review on the Versatile Applications of Ionic Liquids. In Current Innovations in Chemical and Materials Sciences Vol. 4 (pp. 41–57). B P International. https://doi.org/10.9734/bpi/cicms/v4/7049E

Abu Talip, R. A., Yahya, W. Z. N., & Bustam, M. A. (2022). Understanding the physicochemical and transport properties of pyrazolium based ionic liquids bearing iodide and triiodide anions. *Journal of Molecular Liquids, 346*, 118270. https://doi.org/10.1016/j.molliq.2021.118270

Adawiyah, N., Moniruzzaman, M., Hawatulaila, S., & Goto, M. (2016). Ionic liquids as a potential tool for drug delivery systems. *MedChemComm, 7*(10), 1881–1897. https://doi.org/10.1039/C6MD00358C

Al-Sheakh, L., Fritsch, S., Appelhagen, A., Villinger, A., & Ludwig, R. (2022). Thermodynamically Stable Cationic Dimers in Carboxyl-Functionalized Ionic Liquids: The Paradoxical Case of "Anti-Electrostatic" Hydrogen Bonding. *Molecules, 27*(2), 366. https://doi.org/10.3390/molecules27020366

Bernardino, K., Zhang, Y., Ribeiro, M. C. C., & Maginn, E. J. (2020). Effect of alkyl-group flexibility on the melting point of imidazolium-based ionic liquids. *The Journal of Chemical Physics, 153*(4). https://doi.org/10.1063/5.0015992

Bhowmik, D., Kamandar, P., Gupta, J., Mete, S., & Hatvate, N. (2024). An efficient, green and micellar catalyzed preparative-scale synthesis of 3,4-Disubstituted isoxazole-5(4H)-ones in the water. *Sustainable Chemistry for the Environment, 5*, 100070. https://doi.org/10.1016/j.scenv.2024.100070

Brunetti, B., Ciccioli, A., Gigli, G., Lapi, A., Simonetti, G., Toto, E., & Vecchio Ciprioti, S. (2023). Evaporation/Decomposition Behavior of 1-Butyl-3-Methylimidazolium Chloride (BMImCL) Investigated through Effusion and Thermal Analysis Techniques. *Thermo, 3*(2), 248–259. https://doi.org/10.3390/thermo3020015

Chakraborty, P., Sharma, A., Luqman, M., Khan, Mohd. F., & Sankhydhar, K. (2022). CO2-triggered switchable polarity solvents and their advancements. In Green Sustainable Process for Chemical and Environmental Engineering and Science (pp. 149–156). Elsevier. https://doi.org/10.1016/B978-0-12-819850-6.00008-5

Chiappe, C., & Pomelli, C. S. (2014). Point-Functionalization of Ionic Liquids: An Overview of Synthesis and Applications. *European Journal of Organic Chemistry, 2014*(28), 6120–6139. https://doi.org/10.1002/ejoc.201402093

Choudhary, G., Dhariwal, J., Saha, M., Trivedi, S., Banjare, M. K., Kanaoujiya, R., & Behera, K. (2023). Ionic liquids: environmentally sustainable materials for energy conversion and storage applications. *Environmental Science and Pollution Research, 31*(7), 10296–10316. https://doi.org/10.1007/s11356-023-25468-w

Clarke, C. J., Baaqel, H., Matthews, R. P., Chen, Y., Lovelock, K. R. J., Hallett, J. P., & Licence, P. (2022). Halometallate ionic liquids: thermal properties, decomposition pathways, and life cycle considerations. *Green Chemistry, 24*(15), 5800–5812. https://doi.org/10.1039/D2GC01983C

Damodaran, K. (2022). Recent advances in NMR spectroscopy of ionic liquids. *Progress in Nuclear Magnetic Resonance Spectroscopy, 129*, 1–27. https://doi.org/10.1016/j.pnmrs.2021.12.001

David, V., & C. Moldoveanu, S. (2022). Ionic Liquids in Liquid Chromatography. In Analytical Liquid Chromatography - New Perspectives. IntechOpen. https://doi.org/10.5772/intechopen.104122

de Souza, J. P., Pivnic, K., Bazant, M. Z., Urbakh, M., & Kornyshev, A. A. (2022). Structural Forces in Ionic Liquids: The Role of Ionic Size Asymmetry. *The Journal of Physical Chemistry B, 126*(6), 1242–1253. https://doi.org/10.1021/acs.jpcb.1c09441

Dong, L., Xia, J., Song, H., Yang, J., Sun, Y., Li, X., & Zhu, X. (2023). Chiral ionic liquids synthesis and their applications in racemic drug separation and analysis. *Journal of Molecular Liquids, 371*, 120982. https://doi.org/10.1016/j.molliq.2022.120982

Dr. D. Valli Sowbhagyam. (2024). Ionic Liquids as Green Solvents: A Comprehensive Review. *International Research Journal on Advanced Engineering Hub (IRJAEH), 2*(02), 220–224. https://doi.org/10.47392/IRJAEH.2024.0035

Endo, T., Sumida, H., Fujii, K., Takahashi, K., & Kimura, Y. (2020). Heterogeneous Structures of Ionic Liquids as Probed by CO Rotation with Nuclear Magnetic Resonance Relaxation Analysis and Molecular Dynamics Simulations. *The Journal of Physical Chemistry B, 124*(46), 10465–10476. https://doi.org/10.1021/acs.jpcb.0c08030

Endo, T., Sunada, K., Sumida, H., & Kimura, Y. (2022). Origin of low melting point of ionic liquids: dominant role of entropy. *Chemical Science, 13*(25), 7560–7565. https://doi.org/10.1039/D2SC02342C

Fabre, E., & Murshed, S. M. S. (2021). A review of the thermophysical properties and potential of ionic liquids for thermal applications. *Journal of Materials Chemistry A, 9*(29), 15861–15879. https://doi.org/10.1039/D1TA03656D

Fei, Y., Chen, Z., Zhang, J., Yu, M., Kong, J., Wu, Z., Cao, J., & Zhang, J. (2021). Thiazolium-based ionic liquids: Synthesis, characterization and physicochemical properties. *Journal of Molecular Liquids, 342*, 117553. https://doi.org/10.1016/j.molliq.2021.117553

Fritsch, S., & Strassner, T. (2024). Synthesis and physical properties of tunable aryl alkyl ionic liquids based on 1-aryl-4,5-dimethylimidazolium cations. *Beilstein Journal of Organic Chemistry, 20*, 1278–1285. https://doi.org/10.3762/bjoc.20.110

Gardas, R. L., & Coutinho, J. A. P. (2008). A group contribution method for viscosity estimation of ionic liquids. *Fluid Phase Equilibria, 266*(1–2), 195–201. https://doi.org/10.1016/j.fluid.2008.01.021

Gousseva, E., Midgley, S. D., Seymour, J. M., Seidel, R., Grau-Crespo, R., & Lovelock, K. R. J. (2022). Understanding X-ray Photoelectron Spectra of Ionic Liquids: Experiments and Simulations of 1-Butyl-3-methylimidazolium Thiocyanate. *The Journal of Physical Chemistry B, 126*(49), 10500–10509. https://doi.org/10.1021/acs.jpcb.2c06372

Greco, F., Shin, S., Williams, F. J., Heller, B. S. J., Maier, F., & Steinrück, H. (2019). Potential Screening at Electrode/Ionic Liquid Interfaces from In Situ X-ray Photoelectron Spectroscopy. *ChemistryOpen, 8*(12), 1365–1368. https://doi.org/10.1002/open.201900211

Guzmán-Pantoja, J., Ornelas-Robles, N., Hernández-Cortez, J. G., & Martínez-Palou, R. (2025). Synthesis and evaluation of poly(ionic liquids) based on vinyl pyridine as catalysts for the alkylation reaction using a real refinery feedstock. *Journal of Molecular Liquids, 429*, 127651. https://doi.org/10.1016/j.molliq.2025.127651

Hassan, R., Nazir, F., Roosh, M., Qaisar, A., Habib, U., Sajini, A. A., & Iqbal, M. (2022). Synthesis, Characterization, Biological Evaluation, and In Silico Studies of Imidazolium-, Pyridinium-, and Ammonium-Based Ionic Liquids Containing n-Butyl Side Chains. *Molecules, 27*(19), 6650. https://doi.org/10.3390/molecules27196650

Hatvate, Navnath T., Ghodse, S. M., Mundlod, K. N., & Telvekar, V. N. (2020). Metal-Free Synthesis of Pyrimidinone Derivatives via Biginelli Reaction Using Aqueous NaICl2. *Letters in Organic Chemistry, 17*(8), 613–617. https://doi.org/10.2174/1570178617666191126095808

Hatvate, N.T., Ghodse, S. M., & Telvekar, V. N. (2018a). Metal-free synthesis of 2-aminothiadiazoles via TBHP-Mediated oxidative C-S bond formation. *Synthetic Communications, 48*(3), 285–290. https://doi.org/10.1080/00397911.2017.1398330

Hatvate, N.T., Takale, B. S., Ghodse, S. M., & Telvekar, V. N. (2018b). Transition metal free large-scale synthesis of aromatic vinyl chlorides from aromatic vinyl carboxylic acids using bleach. *Tetrahedron Letters, 59*(43), 3892–3894. https://doi.org/10.1016/j.tetlet.2018.09.034

Hayes, R., Warr, G. G., & Atkin, R. (2015). Structure and Nanostructure in Ionic Liquids. *Chemical Reviews, 115*(13), 6357–6426. https://doi.org/10.1021/cr500411q

Hunger, L., Al-Sheakh, L., Zaitsau, D. H., Verevkin, S. P., Appelhagen, A., Villinger, A., & Ludwig, R. (2022). Dissecting Noncovalent Interactions in Carboxyl-Functionalized Ionic Liquids Exhibiting Double and Single Hydrogens Bonds Between Ions of Like Charge. *Chemistry – A European Journal, 28*(46). https://doi.org/10.1002/chem.202200949

Hussain, S., Ali, A., Foorginezhad, S., Chen, Y., & Ji, X. (2025). A comprehensive review on ionic liquids and ionic hybrid materials for CO2 separation. *Separation and Purification Technology, 360*, 130997. https://doi.org/10.1016/j.seppur.2024.130997

Jain, S., Sharma, S., Luqman, M., & Khan, Mohd. F. (2022). Switchable solvents for catalysis. In Green Sustainable Process for Chemical and Environmental Engineering and Science (pp. 201–223). Elsevier. https://doi.org/10.1016/B978-0-12-819850-6.00010-3

Javed, M. N., Batool, H., Khan, M. M., Muhammad, S., Hashmi, I. A., Riaz, K., Bari, A., Henderson, W., & Ali, F. I. (2025). Synthesis, Characterization, and Application of Natural Occurring Hydroxy Functionalized Benzoate based Ionic Liquids as Catalyst for 2-amino-7-hydroxy-4H-chromene Derivatives under Ultrasound Irradiation. *Journal of Molecular Structure, 1336*, 142100. https://doi.org/10.1016/j.molstruc.2025.142100

Khudozhitkov, A. E., Stange, P., Paschek, D., Stepanov, A. G., Kolokolov, D. I., & Ludwig, R. (2022). The Influence of Deuterium Isotope Effects on Structural Rearrangements, Ensemble Equilibria, and Hydrogen Bonding in Protic Ionic Liquids. *ChemPhysChem, 23*(23). https://doi.org/10.1002/cphc.202200557

Labeeb, A., Nour, H., El Malah, T., Ward, A., & Fikry, M. (2024). *Novel Ionic Liquid Crystal as Electrolytic Ion Charges Carrier for Energy Storage Applications.* https://doi.org/10.20944/preprints202407.0681.v1

Lakhani, P., Bhanderi, D., & Modi, C. K. (2024). Silica-supported ionic liquids as versatile catalysts: A case study. *Journal of Molecular Liquids*, *408*, 125306. https://doi.org/10.1016/j.molliq.2024.125306

Liu, N., Liang, J., Lu, Y., Xu, Z., & Liu, H. (2024). Interfacial structures and dissociation reactions of protic ionic liquids on Li metal surface: A combined first-principles and ab initio molecular dynamics investigation. *Chemical Physics Letters*, *836*, 141026. https://doi.org/10.1016/j.cplett.2023.141026

Llaver, M., Fiorentini, E. F., Quintas, P. Y., Oviedo, M. N., Botella Arenas, M. B., & Wuilloud, R. G. (2022). Task-specific ionic liquids: Applications in sample preparation and the chemistry behind their selectivity. *Advances in Sample Preparation*, *1*, 100004. https://doi.org/10.1016/j.sampre.2022.100004

Lu, J.-L., Yan, T., Pan, W.-G., & Wang, L.-W. (2025). Absorption of CO2 by amino acid-functionalized ionic liquids: Mechanisms, properties, and outlook. *Journal of Energy Chemistry*, *104*, 594–608. https://doi.org/10.1016/j.jechem.2025.01.003

Ma, J., Zhu, M., Yang, X., Wang, Y., & Wang, B. (2021). Different cation-anion interaction mechanisms of diamino protic ionic liquids: A density functional theory study. *Chemical Physics Letters*, *774*, 138615. https://doi.org/10.1016/j.cplett.2021.138615

Ma, X., Yu, J., Hu, Y., Texter, J., & Yan, F. (2023). Ionic liquid/poly(ionic liquid)-based electrolytes for lithium batteries. *Industrial Chemistry & Materials*, *1*(1), 39–59. https://doi.org/10.1039/D2IM00051B

Mastellone, G., Marengo, A., Sgorbini, B., Rubiolo, P., & Cagliero, C. (2023). Ionic Liquids in Green Sample Preparation. In Green Sample Preparation Techniques (pp. 179–211). The Royal Society of Chemistry. https://doi.org/10.1039/BK9781839166419-00179

Matsumoto, A. (2024). Solution rheology of poly(ionic liquid)s: current understanding and open questions. *Korea-Australia Rheology Journal*, *36*(4), 319–328. https://doi.org/10.1007/s13367-024-00115-z

McNeice, P., Marr, P. C., & Marr, A. C. (2021). Basic ionic liquids for catalysis: the road to greater stability. *Catalysis Science & Technology*, *11*(3), 726–741. https://doi.org/10.1039/D0CY02274H

Meischein, M., & Ludwig, A. (2021). Upscaling nanoparticle synthesis by sputter deposition in ionic liquids. *Journal of Nanoparticle Research*, *23*(6), 129. https://doi.org/10.1007/s11051-021-05248-8

Morales-Ugarte, J. E., Benayad, A., Santini, C. C., & Bouchet, R. (2019). Electrochemical Impedance Spectroscopy and X-ray Photoelectron Spectroscopy Study of Lithium Metal Surface Aging in Imidazolium-Based Ionic Liquid Electrolytes Performed at Open-Circuit Voltage. *ACS Applied Materials & Interfaces*, *11*(24), 21955–21964. https://doi.org/10.1021/acsami.9b00753

Mousavi, S.-P., Atashrouz, S., Nait Amar, M., Hadavimoghaddam, F., Mohammadi, M.-R., Hemmati-Sarapardeh, A., & Mohaddespour, A. (2021). Modeling surface tension of ionic liquids by chemical structure-intelligence based models. *Journal of Molecular Liquids*, *342*, 116961. https://doi.org/10.1016/j.molliq.2021.116961

Mutelet, F. (Ed.). (2023). *Industrial Applications of Ionic Liquids*. IntechOpen. https://doi.org/10.5772/intechopen.100788

Nasirpour, N., Mohammadpourfard, M., & Zeinali Heris, S. (2020). Ionic liquids: Promising compounds for sustainable chemical processes and applications. *Chemical Engineering Research and Design*, *160*, 264–300. https://doi.org/10.1016/j.cherd.2020.06.006

neamah, A. jaber, Ghareeb, M. M., & Rashid, A. M. (2022). Ionic liquid: A green designer solvent for topical delivery of Nano system. *Maaen Journal for Medical Sciences*, *1*(1). https://doi.org/10.55810/2789-9128.1008

Noorhisham, N. A., Amri, D., Mohamed, A. H., Yahaya, N., Ahmad, N. M., Mohamad, S., Kamaruzaman, S., & Osman, H. (2021). Characterization techniques for analysis of imidazolium-based ionic liquids and application in polymer preparation: A review. *Journal of Molecular Liquids, 326*, 115340. https://doi.org/10.1016/j.molliq.2021.115340

Palumbo, O., Cimini, A., Trequattrini, F., Brubach, J.-B., Roy, P., & Paolone, A. (2020). The infrared spectra of protic ionic liquids: performances of different computational models to predict hydrogen bonds and conformer evolution. *Physical Chemistry Chemical Physics, 22*(14), 7497–7506. https://doi.org/10.1039/D0CP00907E

Paolone, A., Haddad, B., Villemin, D., Boumediene, M., Fetouhi, B., & Assenine, M. A. (2022). Thermal Decomposition, Low Temperature Phase Transitions and Vapor Pressure of Less Common Ionic Liquids Based on the Bis(trifuoromethanesulfonyl)imide Anion. *Materials, 15*(15), 5255. https://doi.org/10.3390/ma15155255

Patil, B. R., Hatvate, N. T., Bari, A. H., Pinjari, D. V, & Pandit, A. B. (2023). Ultrasound-assisted facile and efficient synthesis of novel Benzoxazole derivatives from o-aminocardanol using Indion 190 resin as a reusable catalyst. *Journal of Chemical Sciences, 135*(1), 4. https://doi.org/10.1007/s12039-022-02120-7

Pei, Y., Zhang, Y., Ma, J., Fan, M., Zhang, S., & Wang, J. (2022). Ionic liquids for advanced materials. *Materials Today Nano, 17*, 100159. https://doi.org/10.1016/j.mtnano.2021.100159

Philippi, F., Rauber, D., Kuttich, B., Kraus, T., Kay, C. W. M., Hempelmann, R., Hunt, P. A., & Welton, T. (2020). Ether functionalization, ion conformation and the optimization of macroscopic properties in ionic liquids. *Physical Chemistry Chemical Physics, 22*(40), 23038–23056. https://doi.org/10.1039/D0CP03751F

Prikhod'ko, S. A., Shabalin, A. Yu., Shmakov, M. M., Bardin, V. V., & Adonin, N. Yu. (2020). Ionic liquids with fluorine-containing anions as a new class of functional materials: features of the synthesis, physicochemical properties, and use. *Russian Chemical Bulletin, 69*(1), 17–31. https://doi.org/10.1007/s11172-020-2719-5

Qu, Y., Lan, J., Chen, Y., & Sun, J. (2021). Amino acid ionic liquids as efficient catalysts for CO2 capture and chemical conversion with epoxides under metal/halogen/cocatalyst/solvent-free conditions. *Sustainable Energy & Fuels, 5*(9), 2494–2503. https://doi.org/10.1039/D1SE00060H

Rajapriya Inbaraj, N., Song, S., Chang, R., Fujita, K., & Hayashi, T. (2023). Investigation of Hydration States of Ionic Liquids by Fourier Transform Infrared Absorption Spectroscopy: Relevance to Stabilization of Protein Molecules. *Langmuir, 39*(7), 2558–2568. https://doi.org/10.1021/acs.langmuir.2c02851

Ranjan, P., Yadav, S., Sadique, M. A., Khan, R., Chaurasia, J. P., & Srivastava, A. K. (2021). Functional Ionic Liquids Decorated Carbon Hybrid Nanomaterials for the Electrochemical Biosensors. *Biosensors, 11*(11), 414. https://doi.org/10.3390/bios11110414

Ratti, R. (2014). Ionic Liquids: Synthesis and Applications in Catalysis. *Advances in Chemistry, 2014*, 1–16. https://doi.org/10.1155/2014/729842

Ravula, S., Larm, N. E., Mottaleb, M. A., Heitz, M. P., & Baker, G. A. (2019). Vapor Pressure Mapping of Ionic Liquids and Low-Volatility Fluids Using Graded Isothermal Thermogravimetric Analysis. *ChemEngineering, 3*(2), 42. https://doi.org/10.3390/chemengineering3020042

Ruan, J., Chen, L., Wu, X., Qian, S., Xie, K., Zhang, X., Cheng, H., Song, Z., & Qi, Z. (2025). Hydroxyl-functionalization promoted activity and recovery of ionic liquids in direct dimethyl carbonate synthesis from CO2. *Applied Catalysis B: Environment and Energy, 361*, 124557. https://doi.org/10.1016/j.apcatb.2024.124557

Sharma, M., Mondal, D., Sequeira, R. A., Talsaniya, R. K., Maru, D. A., Moradiya, K., & Prasad, K. (2021). Syntheses and characterization of few bio-ionic liquids comprising of cholinium cation and plant derived carboxylic acids as anions. *Journal of the Indian Chemical Society, 98*(11), 100205. https://doi.org/10.1016/j.jics.2021.100205

Sharma, P., Sharma, S., & Kumar, H. (2024). Introduction to ionic liquids, applications and micellization behaviour in presence of different additives. *Journal of Molecular Liquids, 393*, 123447. https://doi.org/10.1016/j.molliq.2023.123447

Shukla, M. K., Tiwari, H., Verma, R., Dong, W.-L., Azizov, S., Kumar, B., Pandey, S., & Kumar, D. (2023). Role and Recent Advancements of Ionic Liquids in Drug Delivery Systems. *Pharmaceutics, 15*(2), 702. https://doi.org/10.3390/pharmaceutics15020702

Shukla, M., Srivastava, N., & Sah, S. (2011). Interactions and Transitions in Imidazolium Cation Based Ionic Liquids. In Ionic Liquids - Classes and Properties. InTech. https://doi.org/10.5772/23845

Singh, S. K., & Savoy, A. W. (2020). Ionic liquids synthesis and applications: An overview. *Journal of Molecular Liquids, 297*, 112038. https://doi.org/10.1016/j.molliq.2019.112038

Stettner, T., & Balducci, A. (2021). Protic ionic liquids in energy storage devices: past, present and future perspective. *Energy Storage Materials, 40*, 402–414. https://doi.org/10.1016/j.ensm.2021.04.036

Tamilarasan, R., Ganesan, K., Subramani, A., Benazir Ali, L., Alam, M. M., & Mohammed, A. (2023). Synthesis, Characterization, Pharmacogenomics, and Molecular Simulation of Pyridinium Type of Ionic Liquids and Their Applications. *ACS Omega, 8*(4), 4146–4155. https://doi.org/10.1021/acsomega.2c07129

Tanzin, T., Arabi, I. I., Mohinuddin, Md., & Hossain, Md. I. (2023). Ammonium-based ionic liquids: synthesis, characterization, computational approach, molecular docking study and phytotoxicity assessment. *Journal of Molecular Structure, 1273*, 134337. https://doi.org/10.1016/j.molstruc.2022.134337

Tie, J., Mao, Z., Zhang, L., Zhong, Y., & Xu, H. (2023). Strong and Ultratough Ionogel Enabled by Ingenious Combined Ionic Liquids Induced Microphase Separation. *Advanced Functional Materials, 33*(52). https://doi.org/10.1002/adfm.202307367

Tsuzuki, S., & Shinoda, W. (2023). Liquid Structures and Ion Dynamics of Ionic Liquids Viewed from Intermolecular Interactions. *The Chemical Record, 23*(8). https://doi.org/10.1002/tcr.202200272

Wang, M., Zhang, M., Zeng, S., Nie, Y., Li, T., Ren, B., Bai, Y., & Zhang, X. (2023). Effective Absorption of Dichloromethane Using Carboxyl-Functionalized Ionic Liquids. *International Journal of Environmental Research and Public Health, 20*(10), 5787. https://doi.org/10.3390/ijerph20105787

Yalcin, D., Drummond, C. J., & Greaves, T. L. (2020). Solvation properties of protic ionic liquids and molecular solvents. *Physical Chemistry Chemical Physics, 22*(1), 114–128. https://doi.org/10.1039/C9CP05711K

Zanatta, M., Antunes, V. U., Tormena, C. F., Dupont, J., & dos Santos, F. P. (2019). Dealing with supramolecular structure for ionic liquids: a DOSY NMR approach. *Physical Chemistry Chemical Physics, 21*(5), 2567–2571. https://doi.org/10.1039/C8CP07071G

Zhao, M., Wu, B., Lall-Ramnarine, S. I., Ramdihal, J. D., Papacostas, K. A., Fernandez, E. D., Sumner, R. A., Margulis, C. J., Wishart, J. F., & Castner, E. W. (2019). Structural analysis of ionic liquids with symmetric and asymmetric fluorinated anions. *The Journal of Chemical Physics, 151*(7). https://doi.org/10.1063/1.5111643

Zhao, Y., Guo, C., Wu, W., & Tong, J. (2023). Synthesis and characterization of physicochemical properties of two new hydroxyl-functionalized ionic liquids. *The Journal of Chemical Thermodynamics, 176*, 106909. https://doi.org/10.1016/j.jct.2022.106909

Zhu, H., & O'Dell, L. A. (2021). Nuclear magnetic resonance characterization of ionic liquids and organic ionic plastic crystals: common approaches and recent advances. *Chemical Communications, 57*(46), 5609–5625. https://doi.org/10.1039/D1CC02151F

Zhuo, Y., Cheng, H.-L., Zhao, Y.-G., & Cui, H.-R. (2024). Ionic Liquids in Pharmaceutical and Biomedical Applications: A Review. *Pharmaceutics, 16*(1), 151. https://doi.org/10.3390/pharmaceutics16010151

Green Synthesis Methods 2

Abstract

Ionic liquids (ILs) have revolutionized modern synthesis as they are efficient, environmentally friendly media that may be used for a wide range of organic, catalytic, and enzymatic transformations. This chapter discusses their role in catalyzing various reactions like hydrogenation, carbonylation, and hydroformylation, in which the tunable polarity and solvation abilities accelerate rates, selectivities, and stability of the catalysts. In organic synthesis, ILs have been shown to promote important processes, such as nucleophilic and electrophilic substitutions, carbon–carbon coupling, cycloadditions, and multi-component reactions, with better yields and under greener conditions. Stabilizing enzymes, increasing efficiency, and facilitating immobilization for green synthesis are essential to biocatalysis. Their broad electrochemical window, high conductivity, and thermal stability further facilitate photochemical and electrochemical processes.

2.1 Introduction

Synthesis and the extraction of both organic and inorganic substances depend on solvents. Commonly utilized as solvents, VOCs have the benefit of being simple to extract from reaction mixtures. Their extreme volatility, however, puts both human and animal health in serious danger. Irritation, nausea, lightheadedness, and, eventually, more serious consequences like liver, kidney, and nervous system damage can result from exposure, whether it is by skin contact, inhalation, or ocular contact. Many environmental issues brought on by rapid industrial growth have been made worse by insufficient regulatory control and delayed identification. These solvents are volatile, which speeds up the loss of the ozone layer and causes acid rain. Furthermore, the release of untreated solvent waste can cause

H. N. Akolkar et al., *Ionic Liquids*, Synthesis Lectures on Sustainable Development,
https://doi.org/10.1007/978-3-032-02746-7_2

eutrophication in rivers, and its presence in the atmosphere can encourage the development of photochemical smog and ground-level ozone. Green synthesis methods are useful in overcoming such problems. (de Jesus and Maciel Filho 2022).

Green synthesis strategies have experienced significant advancements because of rising market requirements for eco-friendly chemical processes. Traditional synthetic production techniques require VOCs and high energy requirements to produce hazardous waste products. Green chemistry provides a framework to reduce the environmental impact of chemical manufacturing through the use of safer solvents and reagents, renewable feedstocks, energy efficiency, and atom economy (Kurniawan et al. 2021). A wide range of techniques, including the use of ILS, deep eutectic solvents, supercritical fluids, microwave and ultrasound-assisted reactions, biocatalysis, and solvent-free systems, have been added to the definition of green synthesis in recent years. These techniques adhere to sustainability principles and often improve yields, selectivity, and reaction rates (Rubab et al. 2022).

In green synthesis, ILs have become an adaptive and sustainable substitute for traditional organic solvents. They have special properties such as very low vapor pressure, good thermal stability, and superior solvation abilities. They are desirable candidates for environmentally friendly chemical processes because of their qualities, especially when considering green chemistry. They also help to produce greener products by lowering emissions and improving reaction efficiency, in contrast to VOCs, which harm the environment (Younesi-Kordkheili and Pizzi 2022). ILs serve two purposes in the finished product as they simultaneously work as solvents and functional components while enhancing both performance and sustainability. These systems function as dual agents to establish their potential for green synthetic operations through reduced harsh reaction conditions as well as VOCs elimination. Future progress in ecologic technology sectors will receive significant contributions from integrating ILs into green synthesis as researchers develop novel structures and applications (Sasi et al. 2021) (Fig. 2.1).

2.1.1 Catalytic Reactions in Ionic Liquids

Various homogenous and heterogenous catalysts were used in organic chemistry for multiple organic transformations (Ghodse et al. 2018; Hatvate et al. 2018, 2019, 2020; Hatvate and Ghodse 2020; Patil et al. 2023; Bhowmik et al. 2024; Hatvate and Gupta 2024). Yet, ILs have demonstrated significant promise as a catalysis system, increasing the effectiveness of a variety of catalysts, such as enzymes, transition-metal complexes, nanoparticles, and traditional heterogeneous catalysts. Choosing various cations and anions allows control of the properties of their hydrophobicity, ionicity, basicity, and miscibility of reactants and end products. This results in better management of those reactions and ensures products are separated easily. An IL around catalysts can affect its shape, the number of substrate molecules entering it, and the formation of the final product. In addition, using

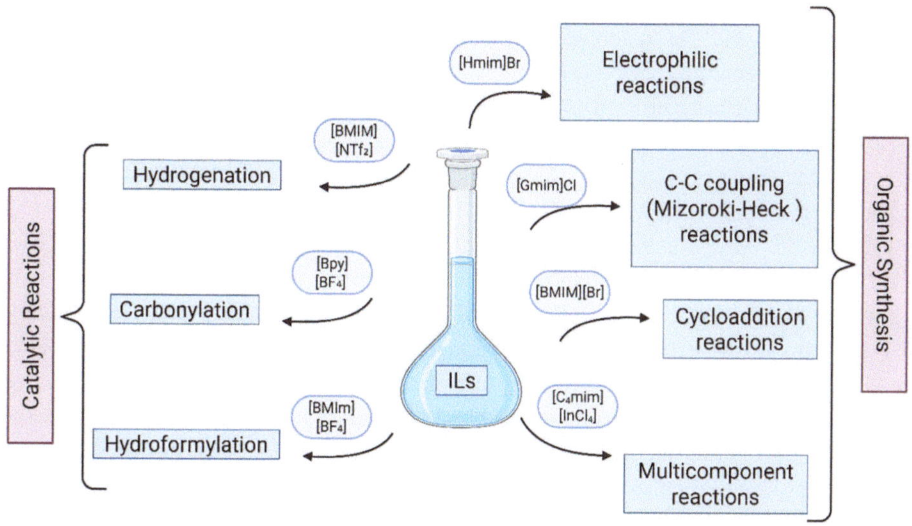

Fig. 2.1 ILs in different reactions

poly(ILs) and supported IL phases makes catalysis more sustainable and allows for the transition to continuous flow. Besides, several products can be separated from catalytic mixtures using decanting, supercritical fluid extraction, or applying temperature and pressure (Migowski et al. 2023). Catalytic hydrogenations, carbonylations, hydroformylations, and aerobic oxidations are catalytic reactions in ILs.

2.1.2 Hydrogenation

ILs have proven to be effective in various hydrogenation processes run more efficiently and selectively. According to Li et al., two bimetallic catalysts, Pd–La, were studied using 1-butyl-3-methylimidazolium acetate as well as 1-vinyl-3-butylimidazolium bromide. They boosted the hydrogenation of 2-ethylanthraquinone and achieved a higher degree of selectivity, making them valuable for hydrogen peroxide production. This improvement modulates the electronic properties of catalytic sites and enhances the interaction between substrate and catalyst (Li et al. 2024a, b). The electronic state of Pt/Pd atoms can be efficiently tuned by ILs, greatly improving their catalytic activity in hydrogenation processes. In particular, they offer a stabilizing environment that encourages a greater concentration of Pd° active species and inhibits Pd nanoparticle agglomeration. Utilizing 1-butyl-3-methylimidazolium hydrogen sulfate for catalyst synthesis, the improved Pd-4% IL catalyst outperforms the traditional Pd/Al_2O_3 catalyst with a remarkable hydrogenation efficiency of 7.7 gL^{-1} and a selectivity of 99.9% toward 2-ethyl-anthraquinone. The potential of IL-assisted systems in creating more effective and selective hydrogenation catalysts is demonstrated by these findings (Wang et al. 2022a, b).

Basic ILs have been utilized in the synthesis of formic acid by providing a buffering environment. 1-Butyl-2,3-dimethylimidazolium acetate is found to be an effective additive in a wide range of solvents. It is used as a medium with an organometallic catalyst such as Ru pincer complex, $Ru_3(CO)_{12}$, RuFe NP, and Pd/C in catalytic hydrogenation of CO_2 to formic acid. Also, for the hydrogenation of CO_2, [Emim][BF_4] is used with Ru/SiO_2 and Cu–ZnO–Al_2O_3 to get methane and methanol, respectively. Such use of ILs provides a good yield of formic acid, methane, and methanol (Sasaki 2022). The combination of ruthenium Ru-MACHO-BH complex and tetrabutylammonium L-argininate enabled the efficient catalytic conversion of CO_2 to methanol. By activating the substrate through hydrogen bonds or ionic interactions, tetrabutylammonium L-argininate most likely makes CO_2 more soluble and facilitates reduction. Under optimal circumstances, this synergistic system produced methanol in good yield and showed significant catalytic activity and selectivity. The process demonstrates how metal catalysts and functional ILs can be used together to create sustainable CO_2 valorization methods (Moazezbarabadi et al. 2025). RuNPs/ILs catalysts demonstrated high selectivity toward the hydrogenation of the vinyl group in styrene (Scheme 2.1), favoring the formation of ethylbenzene over complete saturation to ethylcyclohexane, and successfully hydrogenated styrene to form ethylbenzene and ethylcyclohexane under mild reaction conditions (Krishnan et al. 2023).

The supported IL phase with Wilkinson's catalyst is applied to the hydrogenation of alkenes. 1-butyl-3-methylimidazolium bis(trifluoromethylsulfonyl)imide ([BMIM][NTf_2]) with tris(triphenylphosphine)rhodium(I) chloride used for the preparation of the efficient hydrogenation. As Wilkinson's catalyst is reduced during hydrogenation, it forms charged Rh intermediate species in IL. Like polar organic solvents in SN^1-type reactions, the ionic nature of IL makes them stabilizing agents for charged intermediate states (Kratzer et al. 2023).

ILs enhance the Cu/Al_2O_3 ability to catalyze citral hydrogenation. N-Butyl-N-methylpyrrolidinium trifluoromethanesulfonate, 1-ethyl-2,3-dimethylimidazolium bis(trifluoromethylsulfonyl)imide, and 1-ethyl-3-methylimidazolium bis(trifluoromethylsulfonyl)imide are utilized to coat Cu/Al_2O_3 with a thin film and form a solid catalyst with an IL layer. This catalyst shows improved selectivity for unsaturated alcohols, geraniol, and nerol (Scheme 2.2) while notably inhibiting the production of undesirable isopulegol, which was notable for being synthesized with

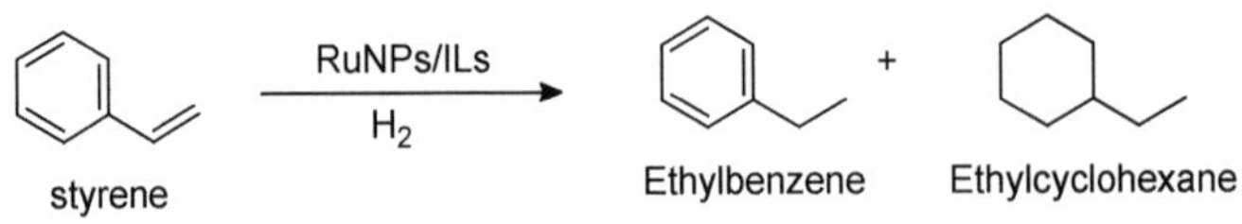

Scheme 2.1 Hydrogenation of styrene

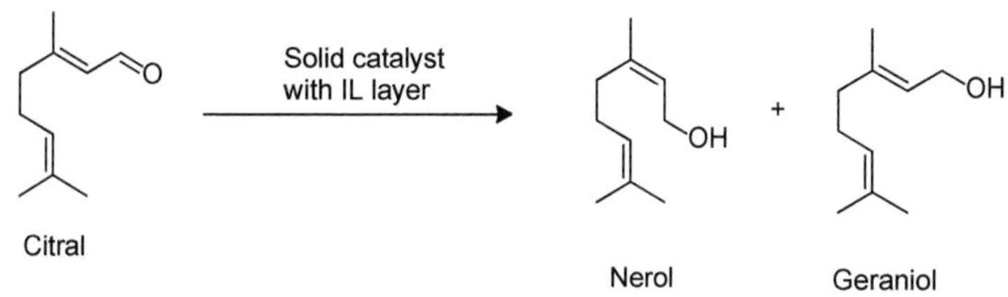

Scheme 2.2 Hydrogenation of citral

Scheme 2.3 Hydrogenation of acetoamidoacrylate

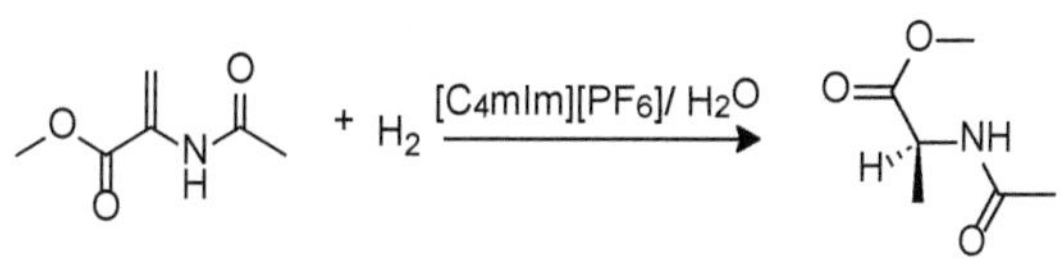

30% selectivity utilizing uncoated catalysts. Among the used ILs, the catalyst coated with 15 wt% 1-ethyl-3-methylimidazolium bis(trifluoromethylsulfonyl)imide achieves 340% improved selectivity for geraniol and nerol compared to the unmodified catalyst. The catalyst surface can be altered by ILs, which can affect reactant adsorption and reaction routes and direct the process toward more desirable molecules (Naicker et al. 2022).

Acetoamidoacrylate (68%) was enantioselectively hydrogenated to N-acetyl-alanine methyl ester (Scheme 2.3) with an excellent enantioselectivity of 96% in a biphasic system consisting of water and $[C_4mIm][PF_6]$. The use of $[C_4mIm][PF_6]$ facilitated the phase separation between catalyst and product and improved chiral induction during hydrogenation (Migowski et al. 2023).

2.1.3 Carbonylation

In catalytic carbonylation reactions, ILs can serve as a solvent and a Lewis acid at the same time, or they can also be a ligand and a solvent simultaneously. Recent developments have shown that Pd-catalyzed amino dicarbonylation of aryl halides can effectively synthesize α-ketoamides (Scheme 2.4). $PdCl_2(MeCN)_2$ in the IL $[Bpy][BF_4]$ coupled with a 3-thiophenyl-benzimidazolium-based diphosphine ligand is a highly efficient system. With respect to α-ketoamides, this catalytic system demonstrated outstanding performance, with high yields of 94–100% and selectivity of 89–95%. It is noteworthy that the catalyst remained active and selective for six runs in a row, highlighting its stability and reusability. In addition to stabilizing the active Pd species, the ionic liquid served as a conducive medium for the dicarbonylation reaction (Árvai and Mika 2024).

Imidazolium ILs are utilized to carbonylate amines with dimethyl carbonate. Ethylenediamine is carbonylated (Scheme 2.5) with dimethyl carbonate in the presence of a

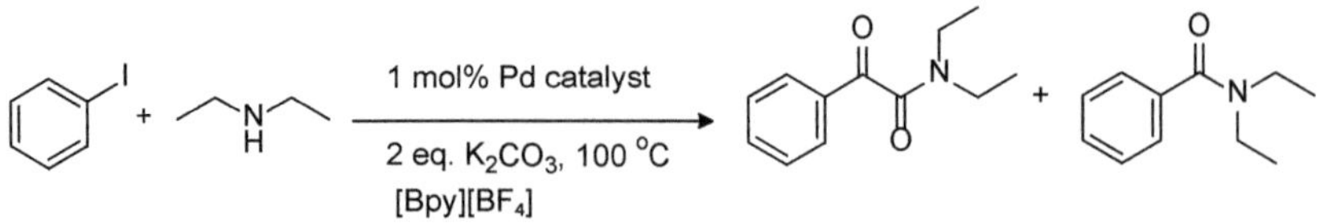

Scheme 2.4 Amino dicarbonylation in IL

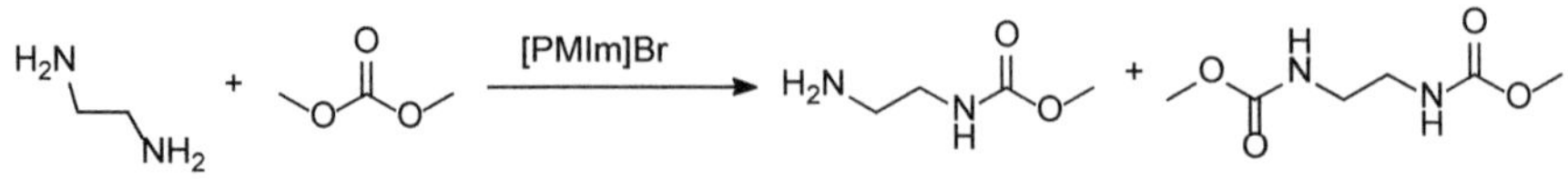

Scheme 2.5 Carbonylation of ethylenediamine

Scheme 2.6 Carbonylation of methanol

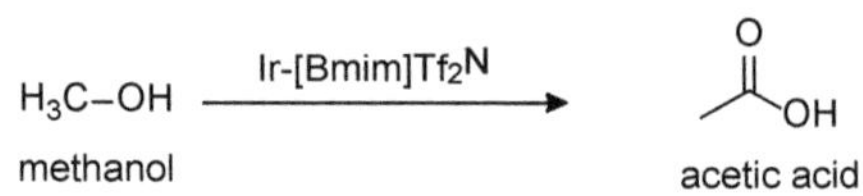

catalyst [PMIm]Br, giving a better yield than in the presence of other catalysts. Dicationic ILs are preferred over monocationic ILs as they give a higher yield (Verma et al. 2021).

Methanol carbonylation to acetic acid (Scheme 2.6) is carried out by utilizing an IL as a non-aqueous liquid phase over Ir(III) catalysts. The catalytic performance of Ir-based systems in the carbonylation of methanol is improved by 1-butyl-3-methylimidazolium bis(trifluoromethanesulfonyl)imide and N-butyl pyridinium bis(trifluoromethanesulfonyl)imide. Under mild conditions (160 °C, 3.0 MPa), these ILs facilitate nearly full methanol conversion and strong acetic acid selectivity. The interaction between the Ir and Tf_2N^- anion forms a stable Ir*[Tf_2N] complex, which is essential to this improved performance. A crucial stage in the carbonylation process, effective CO activation, is possible through this interaction, which also stabilizes the Ir catalyst. Thus, using these ILs presents a viable approach to creating long-lasting and highly effective catalytic systems for methanol carbonylation (Zhang et al. 2022).

2.1.4 Hydroformylation

Alkenes undergo a chemical process called hydroformylation, which involves reacting them with hydrogen and carbon monoxide to produce aldehydes. It is possible to transform homogeneous catalysts into heterogeneous ones by mounting metal complexes on the surface of phosphorus-containing polymeric ILs. This technique combines the ease of separation and reuse of heterogeneous catalysts with the high activity of homogeneous catalysts. The triphenylphosphine ligands can be more precisely tuned by using ILs

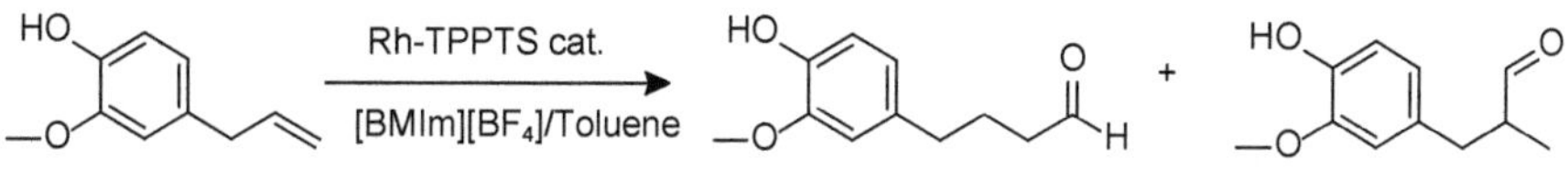

Scheme 2.7 Hydroformylation in [BMIm][BF4]/toluene

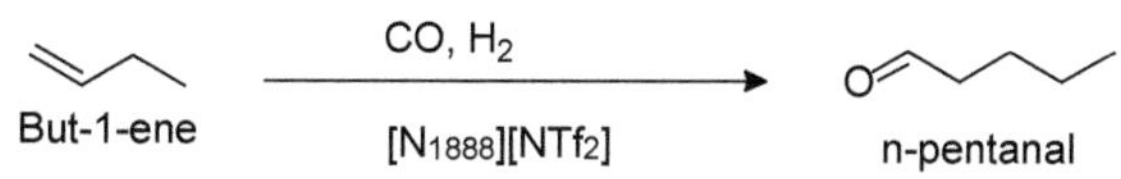

Scheme 2.8 Hydroformylation of but-1-ene

through a copolymerization procedure; a number of PILs-PPh_3 materials were effectively developed, providing reusable catalytic systems (Jin et al. 2022).

Natural olefins such as eugenol, estragole, and safrole were successfully hydroformylated by Rh using a biphasic system consisting of [BMIm][BF_4] and toluene (Scheme 2.7). Under moderate instances (110 °C, 55°bar), the reaction demonstrated high activity and selectivity towards the respective aldehydes using a Rh-triphenylphosphine trisulfonate catalytic system. The toluene phase made it easier to dissolve the substrate and extract the product, while the ionic liquid phase offered a great setting for catalyst solubilization and recycling (Árvai and Mika 2024).

When Rh catalysts are used in a SILP system for the gas-phase hydroformylation of but-1-ene (Scheme 2.8), the performance of the catalyst is influenced by the IL. It demonstrates that catalytic activity is directly impacted by the solubility of reactants, but-1-ene, CO, and H_2, in various ILs. The maximum activity was shown by [N_{1888}][NTf_2] because of its superior olefin solubility. Aldol byproduct generation was also impacted by the specific type of IL anion, with hydrophilic anions limiting the accumulation of aldol byproducts (Schörner et al. 2023).

2.2 Organic Synthesis in ILs

Various organic synthesis reactions, including nucleophilic substitutions, oxidations, reductions, condensations, cycloadditions, and cross-coupling reactions like Suzuki and Heck reactions, are made possible by ILs. They improve reaction rate, yields, and selectivity by acting as catalysts or co-catalysts. They are appropriate for biphasic and multi-component systems due to their capacity to dissolve both organic and inorganic substances. Furthermore, they frequently make it simple to separate and recycle products, which makes organic synthesis more environmentally friendly and sustainable at the commercial level. In the preparation of POFs, such as MOFs and COFs, ILs play a vital role. They facilitate the synthesis of POFs with certain features by acting as solvents, post-synthetic modifiers, and functional components. Their distinct physicochemical properties aid in the formation of frameworks with improved functionality, stability, and porosity.

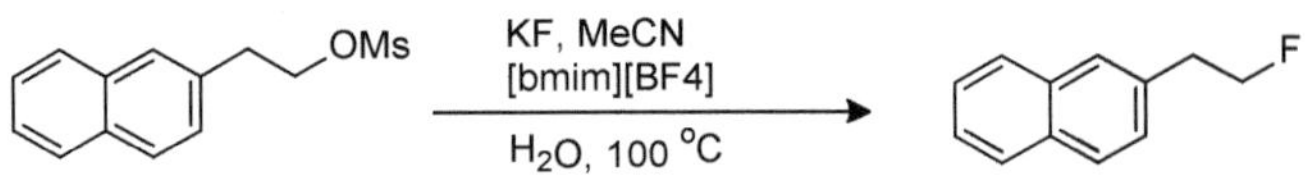

Scheme 2.9 Nucleophilic fluorination using IL

Scheme 2.10 Electrophilic substitution on indole

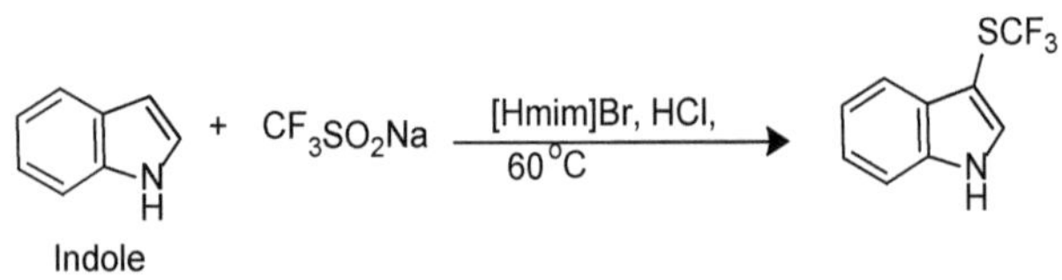

Significant improvements have been made in a number of applications, including gas storage, separation procedures, and catalysis, as a result of the incorporation of ILs into POFs. IL-functionalized POFs have shown enhanced efficiency in gas adsorption and separation tasks, as well as enhanced catalytic performance and selectivity in organic reactions (Li et al. 2023).

2.2.1 Nucleophilic and Electrophilic Reactions

The strong electrostatic interactions between fluoride ions and their counter-cations, which can decrease nucleophilicity, and the limited solubility of metal fluorides in organic solvents are obstacles to nucleophilic fluorination. ILs have a role in enhancing the rate of nucleophilic fluorination. They enhance the solubility of metal fluorides, such as KF and CsF, which promotes improved fluoride-substrate interaction. Because of their ionic nature, they can interact with the substrate and the fluoride ion, stabilizing transition states and reducing activation energy (Oh et al. 2022) (Scheme 2.9).

ILs are recyclable solvents in electrophilic substitution reactions. The environmentally friendly and effective method to add a trifluoromethylthio group to indole compounds (Scheme 2.10) is by using an imidazolium-based IL. There is no requirement for phosphorus compounds or transition metals. 1-Hexyl-3-methylimidazolium bromide ([Hmim]Br) functions both as a solvent and a reductant, and it can be reused afterward, decreasing waste (Wang et al. 2021).

2.2.2 C–C Coupling Reactions

A well-known reaction to form C–C bonds in organic synthesis is the Mizoroki-Heck reaction (Scheme 2.11). Iron is immobilized on imidazolium-based IL to form a heterogeneous catalyst [Gmim]Cl–Fe (III), which carries out the reaction without utilizing Pd and Cu. In mild conditions, this environmentally friendly system effectively promotes the

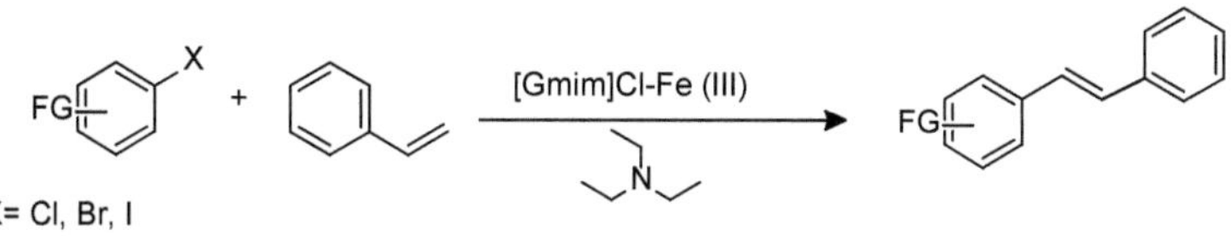

X= Cl, Br, I
FG= H, p-CH_3, p-OCH_3
p-Cl, p-NO_2

Scheme 2.11 Mizoroki-Heck reaction in IL

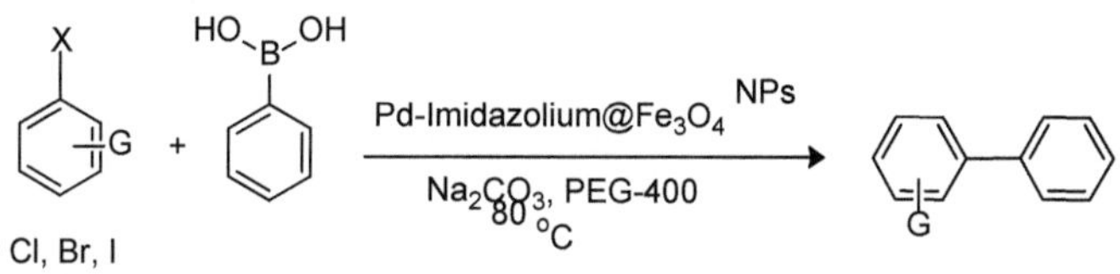

X= Cl, Br, I
G= H, CH_3, NO_2'
OH, Cl, CN

Scheme 2.12 Suzuki cross-coupling reaction in IL

formation of C–C bonds between aryl halides and alkenes. Heterogeneity, chemical and thermal stability, and reusability throughout several catalytic cycles make these systems beneficial (Deepa et al. 2021).

ILs containing Pd have become effective catalytic systems for Suzuki cross-coupling reactions (Scheme 2.12). They are very important for the preparation of organic compounds that have a wide range of structures and are very functional. This particularly enhances nucleopalladation-type Suzuki coupling, which allows C–C bonds to form under moderate and green conditions. A heterogeneous catalyst in which Pd is immobilized on imidazolium-functionalized IL Fe_3O_4 magnetic nanoparticles is developed. This provides a large surface area, better dispersion of Pd active sites, and better stability and recyclability (Imeer et al. 2025).

2.2.3 Cycloaddition Reactions

In cycloaddition reactions, ILs influence regioselectivity and stereoselectivity, favoring endo products. Within the short reaction time, the Diels-Alder reaction (Scheme 2.13) proceeded with good product yields and strong endo-exo selectivity. It was discovered that ILs based on D-glucose and having free −OH groups improve the reactivity and selectivity of the reaction. Specifically, endo/exo ratios were noticeably high for sugar-derived ILs with two free −OH groups (Muthukuru et al. 2021). Using zinc catalysis, cyclic carbonates can be prepared efficiently from carbon dioxide. An efficient catalytic system is made up of $ZnCl_2$ and IL 1-butyl-3-methylimidazolium bromide ([BMIM][Br]), which converts propylene oxide into propylene carbonate. It takes place in mild conditions

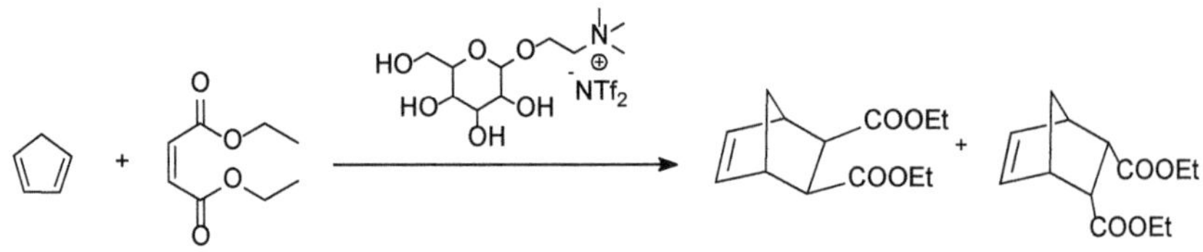

Scheme 2.13 Cycloaddition reaction

and allows the synthesis of up to 95% of the product yield. Here, the $ZnCl_2$ helps open the epoxide ring, while the IL enhances CO_2 solubility and stabilizes the intermediates formed during the reaction (Nasirov et al. 2022).

2.2.4 Multi-component Reactions

ILs have the ability to dissolve diverse reactants and provide a tunable reaction medium. In(III)-based ILs are utilized for the synthesis of pyrazole (Scheme 2.14), specifically in the case of regioselectivity of reaction between phenylhydrazine, benzaldehyde, and 1-phenylbutane-1,3-dione, as only one regioisomer is formed. 1-Butyl-3-methylimidazolium tetrachloroindate ([C_4mim][$InCl_4$]) is applied in the one-pot preparation of pyrazole (Więcławik and Chrobok 2023).

2.3 Biocatalysis and Enzyme-Assisted Synthesis

In green chemistry, biocatalysis and the use of enzymes are now major approaches that replace traditional chemical methods. Using ILs in biocatalysis, they can improve enzyme activity and selectivity as solvents; as support for immobilization, they help stabilize the enzyme structure, and as additives, they modify the surrounding environment of the enzyme to aid in particular reactions. They also improve enzyme-substrate interactions and increase substrate solubility, leading to higher conversion rates. Their versatility in biocatalytic applications is made possible by their use in both homogeneous and multiphase solutions. Advancements seen in the past few years have proven that they help

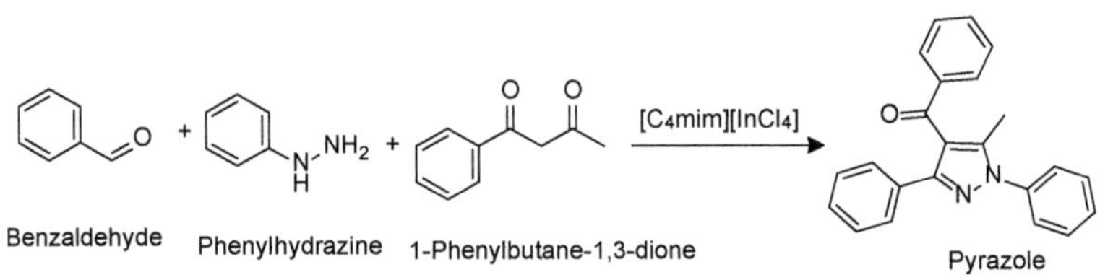

Scheme 2.14 Synthesis of pyrazole

enzymes perform better in biodiesel production, the manufacture of chemicals, and various required materials. This shows that ILs can greatly improve biocatalytic processes and provide a safer and more efficient way of synthesizing chemicals (Elgharbawy et al. 2021; Li et al. 2024a, b).

2.3.1 ILs in Enzyme Activation, Immobilization, and Stability

ILs have been recognized as efficient modifiers for catalysis, activation, and immobilization of enzymes. They are useful in a variety of biotechnological processes because they can stabilize enzymes, increase their activity, and make it easier for them to be reused. Different IL techniques, such as employing them as solvents, supports, and activators, are used to immobilize enzymes. These techniques have been used to increase the catalytic performance and stability of a variety of enzymes, including cellulases and lipases. [C_4mim][BF_4] is used as a solvent for lipase-catalyzed polyester synthesis. A lipase-catalyzed reaction using an IL derivative as the acylating agent in an IL solvent under reduced pressure was used for an effective transesterification system that improved enzyme performance and reaction rate (Badgujar et al. 2022; Itoh 2023). By providing a stable environment that preserves the structure and activity of the enzyme, ILs aid in the more effective operation of enzymes in biocatalytic oxidations. They facilitate access of enzymes to their targets by dissolving a variety of molecules. Furthermore, they are reusable, which lowers waste and improves the sustainability of processes (Sheldon et al. 2023).

The activity of the enzyme β-galactosidase, which is frequently utilized in the food industry, is influenced by certain ILs. They can be used as solvents to either increase or decrease enzyme activity due to their tunability. ILs [emim][BF_4], [bmim][BF_4], and [emim][$MeSO_3$] increased the activity of the enzyme, pointing to the possibility of increased biocatalytic process efficiency. Out of them, [emim][$MeSO_3$] is more effective toward increasing the activity of β-galactosidase. Depending on its structure and characteristics, such as its ability to establish hydrogen bonds or its reactivity, the anion component can influence the stability and activity of an enzyme. Because it can engage with specific sections of the enzyme and generate strong H-bonds, the chloride ion is known to impair enzyme activity by altering its structure. ILs with chloride ions, including [bmim][Cl] and [hmim][Cl], inhibited enzymes like α-amylase. The chloride anion maintains the β-galactosidase activity (Prots et al. 2021) (Fig. 2.2).

2.3.2 IL-Based Biocatalysts in Sustainable Synthesis

A promising type of green media for sustainable chemical synthesis, IL-based biocatalysts provide notable benefits in terms of environmental compatibility, recyclability, and

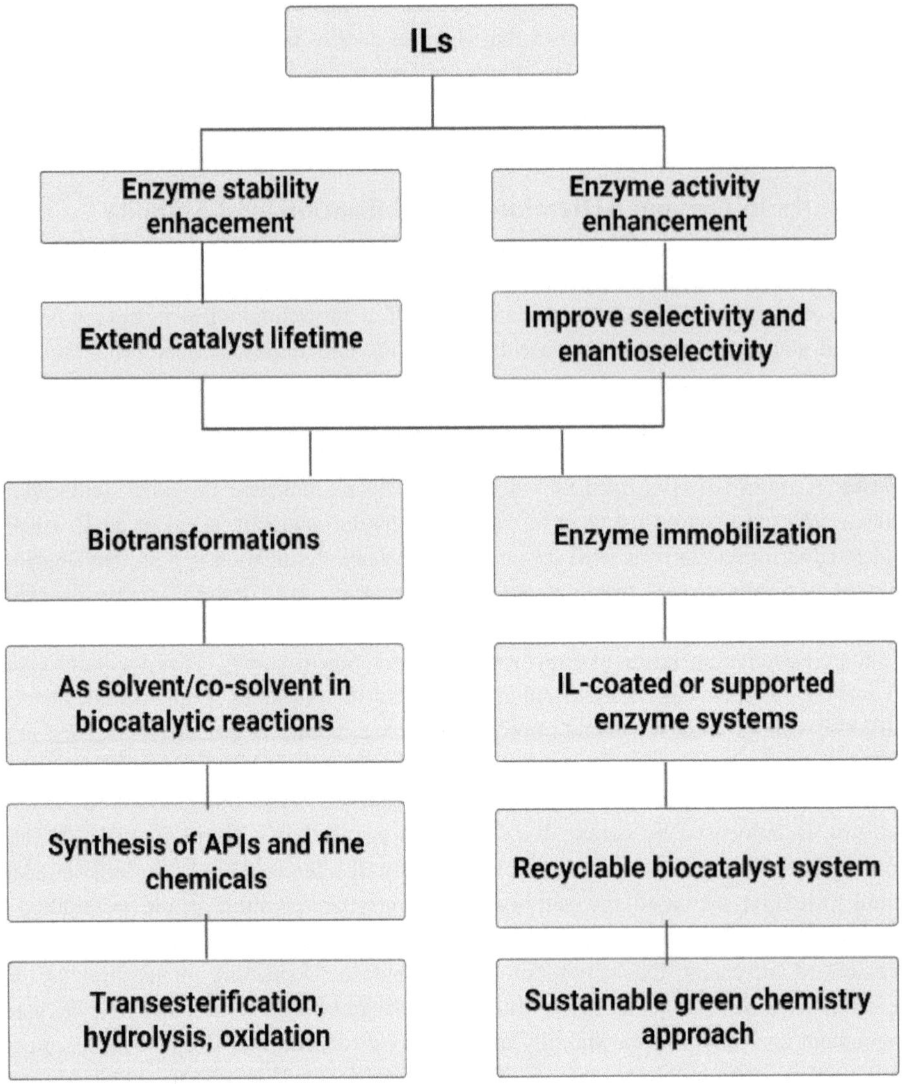

Fig. 2.2 ILs in biocatalysis and enzyme immobilization

enzyme stability. By increasing catalytic efficiency, decreasing waste production, and making product separation simpler, they provide a more environmentally friendly substitute for conventional organic solvents (Badgujar et al. 2022). The Gewald reaction is used to prepare 2-aminothiophenes in an environmentally friendly and effective way. The reaction between 2,5-dihydroxy-1,4-dithiane and several substituted activated nitriles was catalyzed using a biocatalytic system that combined alkali protease lipase, with ILs. The technique maintained steady yields over 10 cycles and produced moderate to outstanding yields with exceptional recyclability. The stability and reusability of the lipase-IL system highlight its potential for useful and sustainable applications in the synthesis of heterocyclic compounds (Zeng et al. 2023). For the synthesis of pure S-ibuprofen with the aid of ILs and a biocatalyst made of lipase enzyme immobilized within hydrogels. By adding lipase enzymes to hydrogels, three-dimensional networks of polymers that can hold water, researchers prepare hydrogel-based biocatalysts. They added ILs with different hydrophobic characteristics to the hydrogel matrix in order to improve the stability and activity of the immobilized enzymes. Several reaction cycles were used to test the stability of immobilized enzymes. Surprisingly, the enzymatic activity remains over 60% after ten cycles, suggesting that the hydrogels supported by IL offered a protective environment that maintained enzyme functionality (Degórska et al. 2024).

A new biocatalyst is formed by fixing lipase, which drives esterification reactions, onto a support made up of magnetic materials, MOFs, and ILs. By incorporating magnetic nanoparticles into a copper-based MOF, a porous structure was formed that gives a large surface for attaching enzymes. Functionalization of the MOF with imidazolium-based ILs containing the $[NTf_2]^-$ was done to increase the activity of the enzyme, working together with the support. They make the environment good for the enzyme, so it remains stable and functions better. Lipase was bound to this IL-functionalized magnetic MOF to ensure it remains strong and prevents leakage in the reaction conditions. Resultant biological catalysts performed much better when used to synthesize isoamyl acetate (Suo et al. 2023). Magnetic nanoparticles immobilize lipase enzymes, which are known to catalyze the transesterification of triglycerides to synthesize biodiesel. Covalent bonds are used to accomplish this immobilization, which guarantees that the enzymes remain bonded to the support material throughout the reaction. A biocatalyst system with excellent catalytic activity, stability, and reusability is produced by combining ILs with magnetic nanoparticle-supported lipase. This system sustains enzyme activity during several reaction cycles and produces notable biodiesel yields (Xing et al. 2023). An innovative green process for biodiesel production utilizes sponge-like ILs, which have properties that make them solid at low temperatures but liquid when heated. When the reaction is performed at 60 °C, sponge-like ILs liquefy, and lipase converts triglycerides into complete biodiesel in a short period. The lipase has a half-life of about 1370 days and is quite stable in this system. The sponge-like ILs solidify when the mixture cools, making it simple to separate the IL from the biodiesel and glycerol. By reusing the enzyme and ionic liquid, this one-pot method effectively integrates extraction, reaction, and separation while promoting sustainability (Lozano et al. 2020).

A study focused on the utilization of ILs to improve lipase immobilization. Both PHBV and silica are combined to prepare hybrid supports, and imidazolium-based ILs called $[C_4mim][NTf_2]$ and $[C_6mim][NTf_2]$ were incorporated during the preparation process. The ILs play a major role in changing the structure and surface properties of the supports, and this is confirmed by BET surface area and FTIR testing. In comparison to supports devoid of them, their presence produced an environment that was more favorable for the enzyme, resulting in a 1.6-fold increase in immobilization yield. Under optimal conditions, these modified supports led to up to 59% conversion from coconut oil to ethyl esters in a transesterification reaction, which proved that ILs can improve the effectiveness of enzyme-based synthesis (Martins et al. 2022).

2.4 Photochemical and Electrochemical Synthesis

In photochemistry and electrochemistry, ILs are safe and useful materials for reactions. By stabilizing reactive intermediates, enhancing catalytic light absorption, and preventing the breakdown of photosensitive compounds, ILs improve the efficiency of photoinduced reactions in photochemical synthesis. Additionally, they allow for the construction of recyclable catalytic systems and selective reactions. Likewise, ILs provide excellent support for redox-active species, high ionic conductivity, and chemical inertness in electrochemical synthesis. These characteristics enable more stable electrodes, less energy consumption, and more efficient electron transfer processes. By lowering solvent toxicity and enhancing reaction selectivity and atom economy, the use of ILs in such synthesis techniques supports the objectives of green chemistry (Dupont et al. 2024).

2.4.1 Photochemical Synthesis in ILs

In photochemical synthesis, reactive intermediates are produced by exposure to light to make the desired products. ILs are good choices for these reactions because of their special properties. ILs are able to stabilize photocatalysts and absorb specific wavelengths of light. They can modulate the electronic properties of catalysts, improving their efficiency in photochemical reactions (Qadir and Dupont 2023). Immobilizing anthraquinone on an IL results in a useful and reusable catalyst for oxidizing vicinal diol groups into benzoic acid derivatives under UV light and air. This combination prevents decomposition and also makes the product simpler to collect. To get the IL-supported anthraquinone, an IL molecule is chemically bonded to the anthraquinone unit. Keeping the photocatalyst in the IL makes it possible to recover and reuse it easily. Because of UV light, anthraquinone transfers an electron to molecular oxygen in the air, forming reactive oxygen species. These species then oxidatively cleave the vicinal diols, yielding benzoic acid derivatives (Koguchi et al. 2023).

Zinc oxide (ZnO) and bismuth oxide (Bi_2O_3) nanoparticles prepared using different ILs, mainly [BMIM]BF_4, [BMIM]PF_6, and [BMIM]Cl, for the effective photodegradation of methylene blue dye in wastewater. At the beginning, they act as solvents and stabilizers to help make the nanoparticles uniform in shape and small in size. Because they have the same structure, the enhanced surface area means that more reactive sites for photocatalysis are formed. ILs used in the fabrication of ZnO and BiO_3 nanoparticles, which greatly increased their photocatalytic efficacy in addition to improving their structural and electrical characteristics (Hussain and Pozan SOYLU 2025). Trihexyl(tetradecyl)phosphonium chloride ($[P_{6,6,6,14}]Cl$) is used to prepare AgCl nanoparticles and magnetic nanocomposites such as $AgCl@Fe_3O_4$ and $TiO_2@Fe_3O_4$. The fact that the synthesis was carried out in a mild environment without the need for high temperatures or other solvents emphasizes the dual function of IL as a solvent and stabilizing agent. For efficient photocatalytic activity, they promoted the production of nanoparticles with regulated sizes and shapes (Corchero et al. 2021).

2.4.2 Electrochemical Synthesis in ILs

ILs have broad electrochemical windows, making them an excellent addition to electrochemical systems. As they can be electrolytes or solvents, they enable reactions to take place under mild conditions, with better selectivity and fewer byproducts. Applications include utilizing CO_2, preparing different chemicals, using metals in electrolysis, and the development of battery materials. They are used as alternatives to typical chemical procedures for making heterocycles, alcohols, and various fine chemicals in the electrochemical process (Wang et al. 2022a, b). Synthesizing arenesulfonyl fluorides directly from nitroarenes can now be achieved proficiently using ILs. One important feature is that N-methylimidazolium *p*-toluenesulfonate ($[Mim]TolSO_3$) acts both as a solvent and an electrolyte, which helps the reaction become more specific. Reactive intermediates are stabilized, and nitroarenes can be reduced much more efficiently at mild electrochemical conditions, all without the need for an extra metal catalyst. The sulfonate anion in IL improves the reaction toward making the sulfonyl fluoride product (Kong et al. 2024). It is possible to prepare controlled nanostructured CuSn alloys with ILs. 1-Butyl-3-methylimidazolium chloride ([BMIM]Cl) combined with $SnCl_2$ and CuCl was selected to prepare a strong and dry electrolyte. This IL combination is appropriate for accurate electrochemical deposition because it has a broad electrochemical window, good ionic conductivity, and thermal stability. A three-electrode setup was used for the electrodeposition step of the synthesis process, which allowed for the preparation of homogeneous CuSn nanowires and macroporous structures at different potentials and temperatures. Due to its intense interaction with metal ions, the IL not only serves as a medium but also has an impact on the nucleation and growth characteristics of alloy (Elbasiony et al. 2022). It was shown that silicon and germanium nanowires can be prepared by depositing them electrically in ILs. In order to provide a broad electrochemical window, high thermal stability, and appropriate solubility for the metal precursors, the process employed 1-butyl-1-methylpyrrolidinium bis(trifluoromethanesulfonyl)imide and 1-hexyl-3-methylimidazolium bis(trifluoromethanesulfonyl)imide as media. They allow easy electrochemical reduction of $SiCl_4$ and $GeCl_4$. The synthesis was conducted utilizing a template-assisted deposition process, in which polystyrene templates were used to guide the production of the nanowires (Lebedeva et al. 2021).

2.5 Conclusion

ILs have been effectively incorporated into a number of green synthesis chemical processes, where they enhance catalyst stability, selectivity, and reaction efficiency. Additionally, their uses in photochemical, electrochemical, and biocatalysis synthesis highlight their versatility. ILs contribute to improved performance and sustainability by serving as catalysts, stabilizers, supports, and solvents. Since their large electrochemical windows, high ionic conductivity, and chemical inertness make them perfect mediums for electron transfer and light-driven processes, ILs have demonstrated special promise in recent developments including photochemical and electrochemical synthesis. They serve as performance enhancers, sustainability enablers, and reaction media. Their incorporation into contemporary green chemistry supports global efforts aimed at producing chemicals in a safer, cleaner, and more sustainable manner. Further advancements in IL design will open up new possibilities for industrial scalability and environmentally responsible synthesis.

References

Árvai, C. and Mika, L.T. (2024) 'Recent Advances in Catalytic Carbonylation Reactions in Alternative Reaction Media', *Chinese Journal of Chemistry*, 42(4), pp. 406–429. Available at: https://doi.org/10.1002/cjoc.202300490

Badgujar, K.C., Badgujar, V.C. and Bhanage, B.M. (2022) 'Recent update on use of ionic liquids for enzyme immobilization, activation, and catalysis: A partnership for sustainability', *Current Opinion in Green and Sustainable Chemistry*, 36, p. 100621. Available at: https://doi.org/10.1016/j.cogsc.2022.100621

Bhowmik, D. *et al.* (2024) 'An efficient, green and micellar catalyzed preparative-scale synthesis of 3,4-Disubstituted isoxazole-5(4H)-ones in the water', *Sustainable Chemistry for the Environment*, 5, p. 100070. Available at: https://doi.org/10.1016/j.scenv.2024.100070

Corchero, R. *et al.* (2021) 'Nanomaterial Synthesis in Ionic Liquids and Their Use on the Photocatalytic Degradation of Emerging Pollutants', *Nanomaterials*, 11(2), p. 411. Available at: https://doi.org/10.3390/nano11020411

Deepa, M. *et al.* (2021) 'Recyclable heterogeneous iron supported on imidazolium ionic liquid catalysed palladium and copper-free Heck reaction', *Journal of Organometallic Chemistry*, 954–955, p. 122073. Available at: https://doi.org/10.1016/j.jorganchem.2021.122073

Degórska, O. *et al.* (2024) 'Ionic liquid supported hydrogel–lipase biocatalytic systems in asymmetric synthesis of enantiomerically pure S-ibuprofen', *International Journal of Biological Macromolecules*, 281, p. 136221. Available at: https://doi.org/10.1016/j.ijbiomac.2024.136221

Jin, F. *et al.* (2022) 'Phosphorus-containing polymeric ionic liquids-supported rhodium catalyst for hydroformylation of dihydrofurans', *Molecular Catalysis*, 521, p. 112207. Available at: https://doi.org/10.1016/j.mcat.2022.112207

Dupont, J. *et al.* (2024) 'Ionic Liquids in Metal, Photo-, Electro-, and (Bio) Catalysis', *Chemical Reviews*, 124(9), pp. 5227–5420. Available at: https://doi.org/10.1021/acs.chemrev.3c00379

Elbasiony, A.M. *et al.* (2022) 'Electrochemical synthesis of nanowires and macroporous CuSn alloy from ionic liquids', *Journal of Solid State Electrochemistry*, 26(3), pp. 783–789. Available at: https://doi.org/10.1007/s10008-021-05109-3

Elgharbawy, A.A.M., Moniruzzaman, M. and Goto, M. (2021) 'Facilitating enzymatic reactions by using ionic liquids: A mini review', *Current Opinion in Green and Sustainable Chemistry*, 27, p. 100406. Available at: https://doi.org/10.1016/j.cogsc.2020.100406

Ghodse, S.M. *et al.* (2018) 'Dual Utility of Heterogeneous Catalyst ZSM-5 for C–C Cleavage Leading to Nitriles, and for the Synthesis of Hydrazides', *ChemistrySelect*, 3(16), pp. 4168–4172. Available at: https://doi.org/10.1002/slct.201800690

Hatvate, N. and Gupta, J. (2024) 'An efficient, green and solvent-free three-component synthesis of 2,4,6-triarylpyridines utilizing ZSM-5 as a heterogeneous catalyst', *Synthetic Communications*, 54(20), pp. 1760–1770. Available at: https://doi.org/10.1080/00397911.2024.2403779

Hatvate, N.T. *et al.* (2020) 'Metal-Free Synthesis of Pyrimidinone Derivatives via Biginelli Reaction Using Aqueous NaICl2', *Letters in Organic Chemistry*, 17(8), pp. 613–617. Available at: https://doi.org/10.2174/1570178617666191126095808

Hatvate, N.T. and Ghodse, S.M. (2020) 'One-pot three-component synthesis of isoxazole using ZSM-5 as a heterogeneous catalyst', *Synthetic Communications*, 50(23), pp. 3676–3683. Available at: https://doi.org/10.1080/00397911.2020.1815786

Hatvate, N.T., Ghodse, S.M. and Telvekar, V.N. (2018) 'Metal-free synthesis of 2-aminothiadiazoles via TBHP-Mediated oxidative C-S bond formation', *Synthetic Communications*, 48(3), pp. 285–290. Available at: https://doi.org/10.1080/00397911.2017.1398330

Hatvate, N.T., Ghodse, S.M. and Telvekar, V.N. (2019) 'ZSM-5 catalyzed one pot three-component synthesis of 5-substituted-1H-tetrazoles from aldehyde', *Synthetic Communications*, 49(24), pp. 3553–3559. Available at: https://doi.org/10.1080/00397911.2019.1679185

Hussain, M.H.A. and Pozan SOYLU, G.S. (2025) 'Synthesis of Ionic Liquid-Assisted Nanoparticles: High Activity, Fast Removal for Photodegradation of Methylene Blue in Water', *Water, Air, & Soil Pollution*, 236(2), p. 97. Available at: https://doi.org/10.1007/s11270-025-07737-1

Imeer, A.T.A. *et al.* (2025) 'Recent Updates on Pd-Incorporated Ionic Liquids Catalyzed Suzuki Cross-Coupling Reaction: A Mini-Review', *Applied Organometallic Chemistry*, 39(2). Available at: https://doi.org/10.1002/aoc.7819

Itoh, T. (2023) 'Enzymatic Reactions using Ionic Liquids for Green Sustainable Chemical Process; Stabilization and Activation of Lipases', *The Chemical Record*, 23(8). Available at: https://doi.org/10.1002/tcr.202200275

de Jesus, S.S. and Maciel Filho, R. (2022) 'Are ionic liquids eco-friendly?', *Renewable and Sustainable Energy Reviews*, 157, p. 112039. Available at: https://doi.org/10.1016/j.rser.2021.112039

Koguchi, S., Fujita, H. and Shibuya, Y. (2023) 'Ionic Liquid-Supported Photocatalysts: A Reusable Environmentally Friendly Oxidation Reaction System That Uses Air and Light', *International Journal of Molecular Sciences*, 24(8), p. 7141. Available at: https://doi.org/10.3390/ijms24087141

Kong, X. *et al.* (2024) 'Direct electrochemical synthesis of arenesulfonyl fluorides from nitroarenes: a dramatic ionic liquid effect', *Green Chemistry*, 26(6), pp. 3435–3440. Available at: https://doi.org/10.1039/D3GC04528E

Kratzer, E. *et al.* (2023) 'Wilkinson-type catalysts in ionic liquids for hydrogenation of small alkenes: understanding and improving catalyst stability', *Catalysis Science & Technology*, 13(7), pp. 2053–2069. Available at: https://doi.org/10.1039/D2CY02058K

Krishnan, D. *et al.* (2023) 'Ruthenium Nanoparticles Stabilized with Methoxy-Functionalized Ionic Liquids: Synthesis and Structure–Performance Relations in Styrene Hydrogenation', *Nanomaterials*, 13(9), p. 1459. Available at: https://doi.org/10.3390/nano13091459

Kurniawan, Y.S. *et al.* (2021) 'Green Chemistry Influences in Organic Synthesis : a Review', *Journal of Multidisciplinary Applied Natural Science*, 1(1), pp. 1–12. Available at: https://doi.org/10.47352/jmans.v1i1.2

Lebedeva, O., Kultin, D. and Kustov, L. (2021) 'Electrochemical Synthesis of Unique Nanomaterials in Ionic Liquids', *Nanomaterials*, 11(12), p. 3270. Available at: https://doi.org/10.3390/nano11123270

Li, W. *et al.* (2023) 'Ionic liquids promoted synthesis, enhanced functions, and expanded applications of porous organic frameworks', *Coordination Chemistry Reviews*, 493, p. 215304. Available at: https://doi.org/10.1016/j.ccr.2023.215304

Li, X. *et al.* (2024a) 'Ionic Liquids Boost Anthraquinone Hydrogenation over Pd-based Bimetallic Pincer Catalysts', *ChemCatChem*, 16(17). Available at: https://doi.org/10.1002/cctc.202400546

Li, Z. *et al.* (2024b) 'Ionic liquids as a tunable solvent and modifier for biocatalysis', *Catalysis Reviews*, 66(2), pp. 484–530. Available at: https://doi.org/10.1080/01614940.2022.2074359

Lozano, P. *et al.* (2020) 'Green biocatalytic synthesis of biodiesel from microalgae in one-pot systems based on sponge-like ionic liquids', *Catalysis Today*, 346, pp. 87–92. Available at: https://doi.org/10.1016/j.cattod.2019.01.073

Martins, S.R.S. *et al.* (2022) 'Ionic liquid synergistic effect between preparation of hybrid supports and immobilization of lipase applied to esters production', *Journal of Thermal Analysis and Calorimetry*, 147(2), pp. 1143–1156. Available at: https://doi.org/10.1007/s10973-020-10408-4

Migowski, P., Lozano, P. and Dupont, J. (2023) 'Imidazolium based ionic liquid-phase green catalytic reactions', *Green Chemistry*, 25(4), pp. 1237–1260. Available at: https://doi.org/10.1039/D2GC04749G

Moazezbarabadi, A. *et al.* (2025) 'Amino Acid-Based Ionic Liquids-Aided CO2 Hydrogenation to Methanol', *ChemSusChem*, 18(7). Available at: https://doi.org/10.1002/cssc.202401813

Muthukuru, P. *et al.* (2021) 'Naturally derived sugar-based ionic liquids: an emerging tool for sustainable organic synthesis and chiral recognition', *New Journal of Chemistry*, 45(43), pp. 20075–20090. Available at: https://doi.org/10.1039/D1NJ03914H

Naicker, L. *et al.* (2022) 'Influencing the Product Distribution in Citral Hydrogenation Using Ionic Liquid Modified Cu Catalysts', *ChemCatChem*, 14(19). Available at: https://doi.org/10.1002/cctc.202200388

Nasirov, F., Nasirli, E. and Ibrahimova, M. (2022) 'Cyclic carbonates synthesis by cycloaddition reaction of CO2 with epoxides in the presence of zinc-containing and ionic liquid catalysts', *Journal of the Iranian Chemical Society*, 19(2), pp. 353–379. Available at: https://doi.org/10.1007/s13738-021-02330-9

Oh, Y.-H., Kim, D.W. and Lee, S. (2022) 'Ionic Liquids as Organocatalysts for Nucleophilic Fluorination: Concepts and Perspectives', *Molecules*, 27(17), p. 5702. Available at: https://doi.org/10.3390/molecules27175702

Patil, B.R. *et al.* (2023) 'Ultrasound-assisted facile and efficient synthesis of novel Benzoxazole derivatives from o-aminocardanol using Indion 190 resin as a reusable catalyst', *Journal of Chemical Sciences*, 135(1), p. 4. Available at: https://doi.org/10.1007/s12039-022-02120-7

Prots, S. *et al.* (2021) 'Added value of ionic liquids in a biocatalytic process: An automatic approach', *Process Biochemistry*, 108, pp. 121–128. Available at: https://doi.org/10.1016/j.procbio.2021.06.003

Qadir, M.I. and Dupont, J. (2023) 'Thermo- and Photocatalytic Activation of CO 2 in Ionic Liquids Nanodomains', *Angewandte Chemie International Edition*, 62(31). Available at: https://doi.org/10.1002/anie.202301497

Rubab, L. *et al.* (2022) 'Green Chemistry in Organic Synthesis: Recent Update on Green Catalytic Approaches in Synthesis of 1,2,4-Thiadiazoles', *Catalysts*, 12(11), p. 1329. Available at: https://doi.org/10.3390/catal12111329

Sasaki, T. (2022) 'CO2 hydrogenation in ionic liquids: Recent update', *Current Opinion in Green and Sustainable Chemistry*, 36, p. 100633. Available at: https://doi.org/10.1016/j.cogsc.2022.100633

Sasi, R. *et al.* (2021) 'Ionic liquid based antimicrobial coating on polymeric surface: A green chemistry approach', *Results in Surfaces and Interfaces*, 5, p. 100026. Available at: https://doi.org/10.1016/j.rsurfi.2021.100026

Schörner, M. *et al.* (2023) 'Tuning catalyst performance in the SILP-catalyzed gas-phase hydroformylation of but-1-ene by choice of the ionic liquid', *Journal of Ionic Liquids*, 3(2), p. 100061. Available at: https://doi.org/10.1016/j.jil.2023.100061

Sheldon, R.A., Bode, M.L. and Mathebula, N. (2023) 'Green and sustainable solvents for biocatalytic oxidations', *Current Opinion in Green and Sustainable Chemistry*, 39, p. 100741. Available at: https://doi.org/10.1016/j.cogsc.2022.100741

Suo, H. *et al.* (2023) 'Covalent immobilization of lipase on an ionic liquid-functionalized magnetic Cu-based metal–organic framework with boosted catalytic performance in flavor ester synthesis', *Journal of Materials Chemistry B*, 11(6), pp. 1302–1311. Available at: https://doi.org/10.1039/D2TB02246J

Verma, K. *et al.* (2021) 'Ionic liquid mediated carbonylation of amines: Selective carbamate synthesis', *Sustainable Chemistry and Pharmacy*, 20, p. 100377. Available at: https://doi.org/10.1016/j.scp.2021.100377

Wang, F. *et al.* (2022) 'Intensifying strategy of ionic liquids for Pd-based catalysts in anthraquinone hydrogenation', *Catalysis Science & Technology*, 12(6), pp. 1766–1776. Available at: https://doi.org/10.1039/D1CY01986D

Wang, F., Lu, G.-P. and Lin, Y. (2021) 'Transition-metal- and phosphorus-free electrophilic trifluoromethylthiolation of indoles with sodium trifluoromethanesulfinates in ionic liquids', *Tetrahedron Letters*, 70, p. 153015. Available at: https://doi.org/10.1016/j.tetlet.2021.153015

Wang, J. *et al.* (2022) 'Challenges and Opportunities of Ionic Liquid Electrolytes for Rechargeable Batteries', *Crystal Growth & Design*, 22(9), pp. 5770–5784. Available at: https://doi.org/10.1021/acs.cgd.2c00706

Więcławik, J. and Chrobok, A. (2023) 'Gallium(III)- and Indium(III)-Containing Ionic Liquids as Highly Active Catalysts in Organic Synthesis', *Molecules*, 28(4), p. 1955. Available at: https://doi.org/10.3390/molecules28041955

Xing, X. *et al.* (2023) 'An efficient and stable magnetic nano-biocatalyst for biodiesel synthesis in recyclable ionic liquids', *Biomass Conversion and Biorefinery*, 13(13), pp. 11947–11957. Available at: https://doi.org/10.1007/s13399-021-02140-2

Younesi-Kordkheili, H. and Pizzi, A. (2022) 'Synthesis of Lignin-Based Polyacid in an Acidic Ionic Liquid: A Green Method to Improve the Performance of Lignin as Catalyst in Urea-Formaldehyde Resin', *Journal of Renewable Materials*, 10(2), pp. 237–246. Available at: https://doi.org/10.32604/jrm.2022.017851

Zeng, Q. *et al.* (2023) 'An efficient synthesis of 2-aminothiophenes via the Gewald reaction catalyzed by enzymes in ionic liquids', *Journal of Molecular Structure*, 1294, p. 136511. Available at: https://doi.org/10.1016/j.molstruc.2023.136511

Zhang, Q. *et al.* (2022) 'Non-Aqueous Liquid Phase Synthesis of Acetic Acid via Ionic Liquid Promoted Homogeneous Carbonylation of Methanol over Ir(III) Catalysts', *Bulletin of the Chemical Society of Japan*, 95(3), pp. 451–457. Available at: https://doi.org/10.1246/bcsj.20210306

3 Applications of Ionic Liquids

Abstract

Ionic liquids (ILs) have attracted many researchers to work on them as an important ingredient in many applications. The chapter concerns the broad uses of ILs in pharmaceutical and diagnostics, extraction technologies, and waste management. They are flexible solvents and catalysts that promote green chemistry and aid in the efficient production of active pharmaceutical ingredients (APIs) in fine chemical and pharmaceutical sectors. In drug delivery, ILs enhance drug-protein binding affinity, enhance the permeability of medicines through biological membranes, and significantly enhance the solubility and bioavailability of poorly water-soluble substances. In medical applications, ILs are aiding the development of high-performance biosensors and electrochemical devices through improvements in sensitivity, signal stability, and biocompatibility. The latter is important with regard to the green extraction of bioactive compounds using microwave-assisted extraction (MAE), ultrasound-assisted extraction (UAE), and solid–liquid extraction (SLE). Also, they are good antistatic agents in polymers and are used to remove pollutants in wastewater, which contributes to sustainability.

3.1 Introduction

ILs are purely ionic and are liquids below 100 °C, which is why they are remarkable with their low evaporation rates, strong resistance to heating, and versatile solvation behavior. This enables their utilization in alternative conventionalities as solvents and catalysts in agreement with the tenets of green chemistry (Neamah et al. 2022). They are being used more often during drug formulation and extraction because of their special solvation properties and biocompatibility. Improved drug delivery methods and the stability of delicate

H. N. Akolkar et al., *Ionic Liquids*, Synthesis Lectures on Sustainable Development,
https://doi.org/10.1007/978-3-032-02746-7_3

biomolecules are made possible by their capacity to dissolve a wide variety of medicinal substances. Besides pharmaceutical applications, they are also indispensable in environmental remediation, particularly in purifying gases and wastewater treatment (Pedro et al. 2021a; Zhuo et al. 2024).

ILs can play an important role in biomass valorization by encouraging the efficient separation of biomass into fermentable sugars, lignin-derived chemicals, and other platform chemicals. They promote effective fractionation under moderate conditions by disrupting the complex network of hydrogen bonding in lignocellulosic biomass that supports the production of sustainable biofuel and biorefinery processes (Rawat et al. 2022). Due to their wide electrochemical stability windows, high ionic conductivity, and non-flammability, they find extensive applications as better electrolytes in batteries and supercapacitors in energy storage applications (Ray and Saruhan 2021). Hence, by increasing efficiency, they play a crucial role in the advancement of green energy technology.

3.2 ILs Pharmaceutical and Fine Chemicals

Recently, there has been much interest in the use of ILs in the pharmaceutical industry because of their potential to solve problems with traditional drug formulations. These difficulties, which are frequently seen in crystalline drugs, include polymorphism, low membrane permeability, chemical instability, and reduced bioavailability. By altering the cation and anion components, they can be developed to enhance pharmacokinetic properties, enable tailored distribution, and boost drug solubility. They serve as solvents, co-solvents, permeability enhancers, and stabilizing agents, offering a possible basis for future pharmaceutical technologies that are more efficient and patient-friendly (Zhuo et al. 2024). Ions like 1-butyl-3-methylimidazolium (C_4MIM) with NTf_2^-, BF_4^-, or PF_6^- are frequently used as typical ILs in the production of APIs. They have demonstrated superior reaction capability owing to being stable at high temperatures, not volatile, and dissolving a broad range of compounds. Even though numerous new cation and anion combinations are being introduced, C_4MIM-based ILs are frequently preferred in pharmaceutical synthesis because they reach better outcomes and higher selectivity (Pedro et al. 2021a).

3.2.1 ILs as a Solvent and Catalyst

Many scientists are paying attention to water, ILs, deep eutectic solvents, and supercritical carbon dioxide for their usefulness in green chemical and pharmaceutical processes (Hatvate et al. 2018a, b, 2020; Patil et al. 2023; Bhowmik et al. 2024; Gulhane et al. 2025). In particular, ILs are remarkable for being stable in both temperature and chemical exposure and for their ability to dissolve substances. Since VOCs are used mainly in pharmaceutical manufacturing and are hazardous, toxic, flammable, and a source of air pollution, they form a major concern for the environment. That is why scientists seek more environmentally friendly and safer alternatives. In recent years, there has been an increase in the number of researchers drawn to ILs, as eco-friendly alternatives to conventional organic solvents in drug preparation (Ehsan Kianfar and Sajjad Mafi 2020; Valli Sowbhagyam 2024).

Many homogenous and heterogenous catalysts were used in the pharmaceutical and special chemical industries (Ghodse et al. 2018; Hatvate et al. 2018b, 2019, 2020; Hatvate and Ghodse 2020, 2024; Parit et al. 2024). Amongst them, ILs have been successfully used in pharmaceutical synthesis for processes like dehydration, oxidation, alkylation, and heterocyclic synthesis. Furthermore, they have shown their versatility as both solvents and catalysts by being used in cycloaddition events to create biologically significant heterocycles. They can catalyze a variety of transformations, with improved reaction rates and selectivity as a result of their capacity to stabilize reactive intermediates. Additionally, their acidity or basicity can be tailored by adding particular functional groups, which increases their suitability for a variety of chemical processes. In pharmaceutical synthesis, the dual function of ILs as solvents and catalysts provides several benefits (Shukla et al. 2023; Singh et al. 2024).

A heterogeneous catalyst effectively promotes the cycloaddition of CO_2 with N-aryl epoxy amines to get 3-aryl-2-oxazolidinones Scheme 3.1. This catalyst is prepared by confining 1-butyl-3-methylimidazolium acetate within the pores of the metal–organic framework (MOF) Fe-MIL-101-NH_2. Besides presenting dual functionality as a solvent and catalyst, the introduction of IL into the MOF framework also enhances the activation of both CO_2 and epoxide substrates via synergistic effects. It also offers powerful, efficient, and recyclable catalysts to green chemical transformations. This catalyst can be used to prepare an antidepressant drug, toloxatone (Niu et al. 2024).

Tetracationic ILs, dodecylimidazolium acetate and hexadecylimidazolium acetate are applied as capping agents and solvents in the synthesis of calcium carbonate $CaCO_3$ nanoparticles. In order to achieve the controlled $CaCO_3$ nanoparticle nucleation and growth, tetra-cationic imidazolium IL as a capping agent binds to both calcium ions, Ca^{2+}, and carbonate ions, $CO_3{}^{2-}$. Their nature is amphiphilic, promoting the prevention of agglomeration and stabilization of the nanoparticles (Atta et al. 2025). The

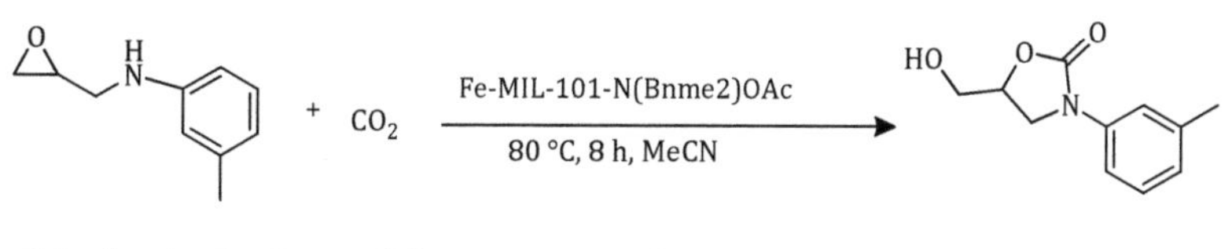

Scheme 3.1 Synthesis of oxazolidinone compound

2-hydroxypyridine-based protic IL is employed in the conversion of CO2 to quinazoline-2,4(1*H*,3*H*)-diones, which is a one-step neutralization reaction. The IL used is 1,1,3,3-tetramethylguanidinium 2-hydroxypyridine in the conversion of 2-aminobenzonitrile to quinazoline-2,4(1*H*,3*H*)-dione Scheme 3.2 (Jiang et al. 2023).

ILs are involved in the heterocyclic synthesis of pyrroles, imidazoles, benzimidazole, and triazoles. They are used both as catalysts to accelerate reaction rates and as solvents to dissolve wide-ranging reactants. They enhance the rates and selectivity of reactions by stabilizing transition states and charged intermediates. Relative to traditional solvents, reactions run in bridged *N*-based ILs often occur under mild conditions, delivering energy efficiency. Benzimidazole can be prepared by a one-pot reaction of benzaldehyde and o-phenylenediamine in the presence of diazabicyclo[2.2.2]octane diacetate ([DABCO][diAc]), which is an N-bridged IL Scheme 3.3 (Chaudhary and Srivastava 2025). Brønsted acidic protic ILs (PILs) were prepared by mixing sulfuric acid and easily available bases. It can be used under mild conditions to esterify levulinic acid with various alcohols, both as a catalyst and solvent in a biphasic system. They found that the acidity of PIL has a direct influence on their catalytic activity, and reaction conditions could be developed to give the highest yields. These ILs are recyclable and have a high yield of up to 99%, and a constant conversion rate and selectivity over 10 cycles (Przypis et al. 2020).

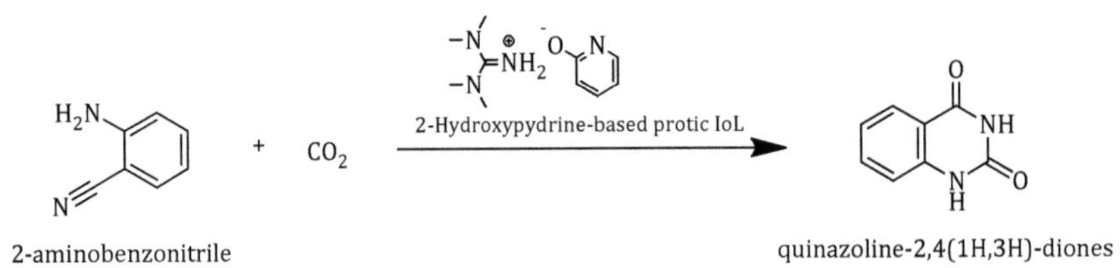

Scheme 3.2 Synthesis of quinazoline-2,4(1H,3H)-dione

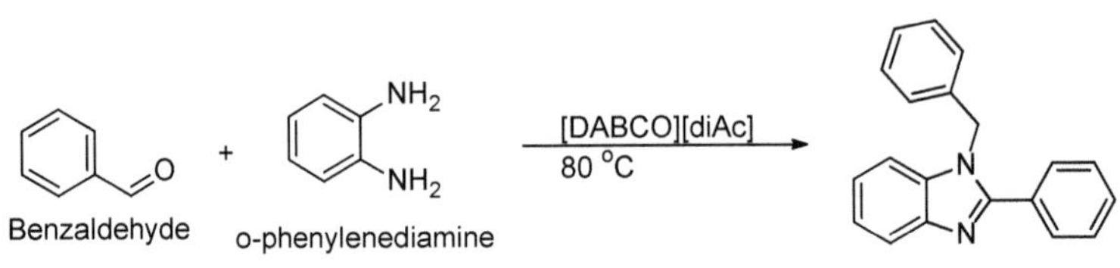

Scheme 3.3 Synthesis of benzimidazole

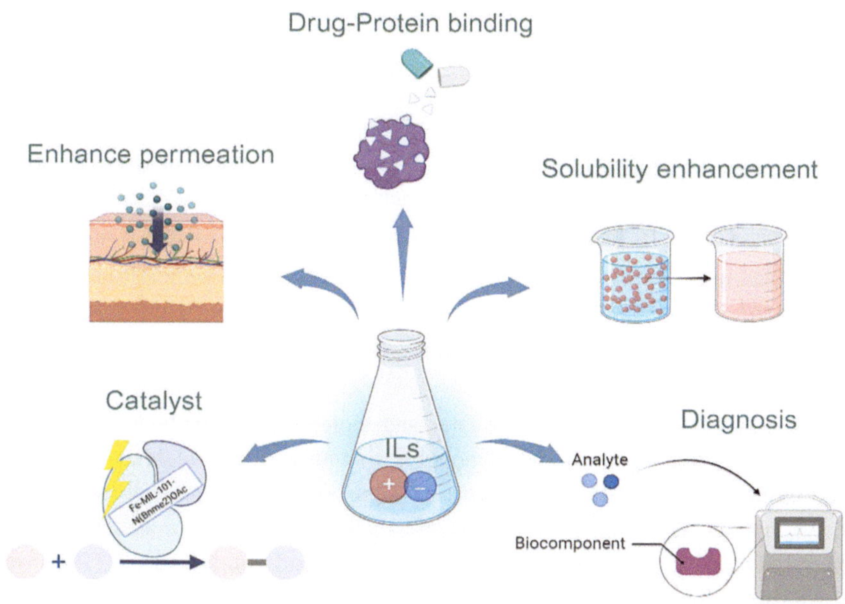

Fig. 3.1 Pharmaceutical applications of ILs

3.2.2 ILs in Drug Delivery

ILs enhance therapeutic performance, safety, and bioavailability; they can be used as antibacterial agents, skin penetration enhancers, macromolecular stabilizers, and effective solubilizers of poorly water-soluble drugs. ILs based on choline, especially, have attracted special attention for use in drug delivery systems. They are extracted based on naturally occurring choline and are suitable for use by humans due to their low or non-toxicity, biocompatibility, and biodegradability. Choline ILs have the ability to facilitate drug transportation across biological membranes, especially in transdermal and oral administration, macromolecule (proteins and peptides) stabilization, and enhancing the solubility of hydrophobic drugs. They also have a role as an antibacterial, with formulation preservation and infection prevention (Li et al. 2022, 2023) (Fig. 3.1 and Table 3.1).

3.2.3 Enhance the Permeability of Drugs

Hydrophobic ILs can insert within the lipid bilayer and disrupt transcellular transport across the membrane, and drugs permeate more efficiently. Hydrophilic ILs can tend to promote paracellular transport by enhancing fluidization and opening of tight junctions in the stratum corneum (Kuddushi et al. 2024). In transdermal drug delivery, ILs are

Table 3.1 ILs in the improvement of drug/protein properties

IL	Drug/protein	Improvement in	References
Choline-fatty acid	Insulin	Macromolecular stabilization	Islam et al. (2021)
Choline prolinate	Zafirlukast	Solubility enhancement	Gaikwad et al. (2022)
Imidazole-based alkyl sulfate	Human serum albumin	Drug-protein binding	Reddy et al. (2018)
β-alanine ethyl ester-favipiravir IL	Favipiravir	Oral delivery	Moshikur et al. (2021)

emerging as novel tools in providing safer and more efficient methods of drug administration through the skin. It is also possible to improve the solubility, stability, and release profiles of medications by preparing them as API-ILs, ionically bonded forms of active ingredients. Under some conditions, they can even behave as smart polymers, forming micelles or gels which enhance distribution and control release. For instance, fatty acid-based ILs transformed hydrophilic medications into oil-soluble forms appropriate for patches, while choline-geranic acid ILs have effectively administered nanosensors (Lu et al. 2022). Choline-geranic acid, a biocompatible IL utilized to transfer nanosensor particles through the skin for ongoing thrombosis monitoring. It is typically necessary to inject thrombosis-detecting nanoparticles, which contain thrombin-sensitive peptide reporters. However, these nanosensors penetrate the dermis and enter the bloodstream over a period of around 72 h when combined with choline geranic acid and applied to the skin (Bekdemir et al. 2022). A study focused on transdermal delivery of insulin, which utilizes ILs for better bioavailability to treat diabetes. A novel **biocompatible IL-based microemulsion was prepared to deliver insulin transdermally. For this formulation, choline propionate is utilized as an internal phase and oil phase of** isopropyl myristate. IL-in-oil microemulsion prepared by combining Span-20 as a cosurfactant with choline-fatty acid as surface-active ILs. Stable, nanoscale droplets were confirmed by TEM and light scattering characterization. By altering the lipid composition of the stratum corneum, these ILs increased fluidity and made it possible for insulin to enter the body through the intercellular pathway (Islam et al. 2021).

A study reveals that the topical delivery of dextran is possible with the help of ILs, which significantly enhances its permeation into the upper dermis up to a depth of 600 μm. Choline-malic acid delivers dextran to both the epidermis and dermis due to the hydrophilicity of malic acid. Skin hydration aids IL in dissociating into ions and allows the upper dermis to retain dextran (Wu et al. 2019). The dissolution and penetration capacity of puerarin and ferulic acid through the skin is enhanced by the performance of choline-amino acid-based ILs. They also facilitate the absorption of drugs into the lower layers of skin by modifying the stratum corneum structure of the epidermis. Choline-glycine IL showed the most significant improvement effect among the tested ILs since it has a strong solubilizing capacity and biocompatibility, indicating that they are safer enhancers of transdermal drug delivery (Yuan et al. 2022).

3.2.4 Enhance Drug-Protein Binding

Drug delivery systems depend on the interaction between drugs and proteins, and ILs have been shown to enhance this binding. Their structural characteristics, including the anionic and cationic components and the length of the alkyl chain, can influence how they interact with proteins. This interaction could lead to improved drug-protein complex stability and binding affinity for drug delivery applications (Pedro et al. 2021b). The structural study of the receptor-binding domain (RBD) of the SARS-CoV-2 spike protein revealed that ILs enhance drug-protein interactions by engaging in special interactions with proteins. This shows that the cationic chain length and its compatibility with the protein influence their affinity, which is that longer alkyl chains bind more strongly to the RBD. Hydrophobic and π-π interactions largely contribute to the high binding between ILs and the RBD that can be used in therapeutic design and development (Liu et al. 2023a, b). Protein binding is enhanced by the anionic constituent of imidazole-based alkyl sulfate ILs, which is evidenced by the binding capacity of these ILs with proteins, such as collagen and human serum albumin. A study shows that the longer the alkyl chain in the anion, the greater the binding affinity between ILs and proteins. Based on this trend, strong binding is favored because it increases hydrophobic interactions. This implies that ILs can be developed to enhance bioavailability and therapeutic effect (Reddy et al. 2018).

3.2.5 Enhance the Solubility of Drugs

ILs show potential as pharmaceutical formulations because they feature distinctive interactions, polarity, and H-bonding properties, which enhance the solubility of low-soluble drugs and raise bioavailability (Bagheri et al. 2024). A study on the synthesis of antiviral drug favipiravir has low water solubility, which limits its oral administration and performance. As a way to overcome this, scientists synthesized new prodrug salts by mixing favipiravir with various organic cations, such as alanine ethyl ester, to make it IL. In order to confirm their structure and stability, NMR, FT-IR, PXRD, TGA, DTG, and DSC techniques were fully applied to characterize this modified favipiravir-IL. Compared with unmodified favipiravir, the favipiravir-IL had a higher solubility in aqueous solutions. (Moshikur et al. 2021). PIL 2-hydroxyethylammonium carboxylate has an excellent effect on the solubility of poorly soluble drugs such as indomethacin, which was increased by up to 200-fold as compared to water because of the strong H-bonding interactions

(Akbarzadeh Gondoghdi et al. 2024). Moreover, through this kind of interaction, zafirlukast solubility can be increased via choline prolinate IL, and thus, it is a probable functional excipient solvent (Gaikwad et al. 2022).

3.3 ILs in Diagnosis

ILs act as customized solvents used to purify diagnostic molecules (proteins, nucleic acids, and metabolites) of biological samples. Their electrochemical stability and flexible structure enhance the performance of diagnostic platforms by improving the clarity, sensitivity, and selectivity of the signal in imaging assays and biosensors (Hu et al. 2023). Synthesis of metal nanoparticles in cancer application, ILs used as a stabilizing agent and a liquid medium. The metal nanostructures are prepared in ILs, and after their formation, the nanoparticles are stabilized and encapsulated by the same IL, and thus they do not agglomerate. The synthesized nanoparticles can then be utilized in imaging and treatment through the appropriate modification in the composition of IL and reaction conditions (Hassanpour et al. 2022).

ILs play an important role in improving the performance of an electrochemical sensor designed to detect α-fetoprotein, which is an essential marker of liver cancer. ILs modify the hydroxyapatite@nano-TiO_2 composite and enhance the stability of the sensing material, electrical conductivity, and dispersion of the nanoparticles. They also aid in achieving a more biocompatible immobilization environment of the antibodies or recognition elements to improve the kinetics of the electron transfer and the effectiveness of binding at the electrode surface. This, compared to IL-free approaches, results in increased sensitivity and selectivity of alpha-fetoprotein detection. The IL also contributes to a more efficient electrochemical response compared to the IL, which serves only as a conductive additive because it can act by structural modification (Ghorbanizamani et al. 2024).

IL-based aqueous biphasic systems are an effective way to extract and purify prostate-specific antigen, a vital biomarker for the detection of prostate cancer. ILs are useful phase-forming elements in this system that help with the selective extraction of prostate-specific antigen from biological fluids like serum because of their low volatility, biocompatibility, and variable polarity. The IL-based method improves the sensitivity, recovery rate, and purity of antigen detection in subsequent immunoassays by lowering interference from other proteins and contaminants. Unlike traditional organic solvents, ILs help provide more precise and eco-friendly diagnostic systems in addition to enhancing biomolecule partitioning and stabilization. A promising technique for high-precision diagnostics is the IL-based aqueous biphasic system, which can be used with a variety of biomarkers (Flora et al. 2023).

IL-based sensors offer a potential approach to diagnosing gout by enabling the real-time, very sensitive detection of uric acid in human blood and urine, which is crucial for an early diagnosis. An innovative electrochemical sensor is developed by modifying the electrode surface with benzimidazolium-1-acetate. The stability and electrical conductivity of the sensor were enhanced by the high π-electron density in IL aromatic rings and carboxyl groups. The sensor showed good sensitivity and selectivity, along with outstanding resistance to interference from other chemicals. Its low detection limit (1.27 μM) and wide linear detection range (2–1050 μM) made it suitable for monitoring uric acid levels in both gout sufferers and healthy individuals. This makes the sensor a practical tool for accurately and quickly detecting hyperuricemia and gout (Abbas et al. 2020; Hu et al. 2023).

3.4 ILs in the Extraction of Bioactive Compounds

ILs are utilized as eco-friendly solvents to extract bioactive chemicals from natural sources. Their strong solvation ability, low volatility, and tunability make them suitable for extracting polyphenols, flavonoids, alkaloids, and essential oils. Task-specific ILs can enhance target molecule stability, selectivity, and extraction efficiency in comparison to traditional solvents (Ferreira et al. 2023) (Table 3.2).

3.4.1 ILs in Microwave-Assisted Extraction (MAE)

ILs, particularly imidazolium-based ILs, enhance chemical solubility and microwave absorption, which boosts the efficiency of MAE. This combination has been shown in numerous studies to enhance the extraction of components such as flavonoids and rosmarinic acid. In contrast to maceration, $[C_4mim][BF_4]$ yielded more polyphenols from *Peperomia pellucida*, whereas $[C_8mim][Br]$ successfully extracted flavonoids from *Scutellaria baicalensis*. Due to their ability to form H-bonds, PILs are utilized for hydrophilic compounds (Lim et al. 2022). The essential oil from *Citrus sinensis* was effectively separated using 1-ethyl-3-methylimidazolium acetate. Enhancing extraction efficiency and

Table 3.2 ILs in extraction of bioactive compounds

Extraction technique	Plant/source	Target compound	IL involved	References
MAE	*Peperomia pellucida*	Polyphenol	$[C_4mim][BF_4]$	Lim et al. (2022)
	Scutellaria baicalensis	Flavonoid	$[C_8mim][Br]$	
UAE	*Chestnut shells*	Pigment	[Bmim]OAc	Zhu et al. (2022)
SLE	*Haematococcus pluvialis*	Astaxanthin	$[BDMC][CO_2]$	Khoo et al. (2021)

raising the quantity of oxyterpene components in essential oil. ILs speed up the extraction process (Franco-Vega et al. 2021).

3.4.2 ILs in Ultrasonic Assisted Extraction (UAE)

Lycopene, a significant antioxidant, is efficiently extracted from guava by UAE with ILs. According to a study, imidazolium-based ILs enhance lycopene extraction by improving cell wall disruption and solubility during ultrasound treatment, making them efficient and eco-friendly solvents. The extraction conditions were optimized using response surface methodology (RSM), and they were further refined using artificial neural network–genetic algorithm (ANN–GA) models. Compared to using traditional solvents, this produced a higher yield (Wang et al. 2024). UAE can be used to extract oil from argan seeds using n-decane and IL 1-propyl-3-methylimidazolium bis(trifluoromethylsulfonyl)imide as solvents. Although IL increases oil output, it cannot be used as an extraction solvent on its own (Hayyan et al. 2025). In the UAE, 1-Butyl-3-methylimidazolium acetate ([Bmim]OAc) works well for efficiently extracting pigment for hair dye from chestnut shells (Zhu et al. 2022).

3.4.3 ILs in Solid–Liquid Extraction (SLE)

SLE is a method of purification that uses a liquid phase and a solid sorbent. ILs have the capacity to make this strategy better. Their ion-exchange properties, π-π interactions, and adaptable polarity improve extraction efficiency and selectivity. They can be employed as solvents for elution or as sorbent modifiers (Feng et al. 2020). SLE was used to extract the antioxidant astaxanthin from the microalga *Haematococcus pluvialis*. The alkyl carbamate ILs used in this instance were butyl dimethyl carbamate ([BDMC][CO_2]) and ethyl dimethyl carbamate. They enhance the release of intracellular astaxanthin and efficiently break down the algal cell walls. The study found that [BDMC][CO_2] had a greater extraction efficiency due to its increased hydrophobicity, which improved the solubilization of astaxanthin (Khoo et al. 2021).

3.5 ILs as an Antistatic Agent

In the packaging and electronics industries, in particular, static charge accumulation in polymers can result in equipment failure, dust attraction, and safety hazards. Over time, typical antistatic chemicals may lose their effectiveness due to migration or limited temperature stability. In polymer matrices, ILs can form percolation networks that reduce surface resistivity and promote charge dissipation. IL/polymer composites can have their conductivity modified by adjusting their structure, concentration, and interaction with the polymer matrix. Imidazolium-based ILs, in particular, have shown promising results in improving the antistatic properties of polyethylene, polystyrene, and polyvinyl chloride (Kosiński et al. 2022). An investigation was conducted to enhance the antistatic qualities of nylon 6 by synthesizing PIL based on quaternary ammonium salts. This forms ionic groups that, either applied directly to the fiber surface or combined with the polymer matrix, dissipate static charge. The quaternary ammonium cation ILs anchored in a polymer chain offer a non-migrating, long-lasting antistatic layer that overcomes the limitations of traditional migrating additives (Fu et al. 2024).

3.6 ILs in Waste Management

Using techniques like solvent extraction, adsorption, and micro-extraction and utilizing interactions like hydrogen bonding and electrostatic forces, ILs have been effectively employed in wastewater treatment to eliminate a range of contaminants, including dyes, organics, heavy metals, and medicines (Goutham et al. 2022). In general, wastewater contains dyes, amines, toluene, herbicides, pesticides, and phenolic compounds that mostly come from the textiles, pharmaceuticals, and agricultural industries. In certain instances, biological treatment and advanced oxidation methods have challenges. It takes longer to use biological methods, and their results can depend a lot on the composition of the wastewater, whereas oxidation processes might result in toxic substances being produced. For this reason, there must be more attention paid to rapid extractive methods because they are both flexible and efficient. However, the choice of solvent, which must have properties like low viscosity, low vapor pressure, and suitable miscibility, greatly influences how effective these processes are. ILs, in particular functionalized ILs, have become viable substitutes for organic pollutants in this regard (Isosaari et al. 2019; Khraisheh et al. 2021).

Liquid–liquid extraction is the main method used to recover organic contaminants from industrial effluent on a large scale. In this case, hydrophobic ILs act as solvents since they can separate into two immiscible layers with water. The organic pollutants in the feed help the hydrophobic IL drop to the bottom because it is denser compared to the aqueous phase. The separation makes it simple to get the IL phase and thus remove the organic pollutants (Padinhattath et al. 2025). Task-specific ILs are tunable and more effective than traditional solvents at removing specific pollutants. This method lowers the production of secondary waste while simultaneously improving the effectiveness of pollution removal (Al Hassan et al. 2023). The liquid membrane extraction technique uses IL mounted on a thin film to achieve separation in a single unit. This covers the removal of pollutants from the feed, their subsequent recovery during the stripping phase, and their diffusion over the membrane. The performance of this process depends on the support material used, as IL Aliquat 336 supported on polypropylene recovers 68% of acetaminophen, while just 29%

on polyvinylidene difluoride. Problems like membrane breakdown during agitation and IL leakage into the aqueous phase persist even though membrane systems are more effective in using IL because of immobilization (Albasri and Wahab 2020; Khraisheh et al. 2021).

As urbanization and industrial activity increase, there is a greater demand for extracting metals from wastewater. Because Ni, Cd, Hg, and Pb cause serious threats to both water bodies and people by easily accumulating and polluting the water, they are subject to stringent regulations. There is special concern about wastewater from hospitals because it can contain the cancer drug platinum, for example, Cisplatin and Oxaliplatin, which are difficult to separate using typical wastewater treatments. As ILs have a tunable structure, they have the capacity to form complexes with metal ions and become attractive extractants for the removal of metals (Roque-Diaz et al. 2021; Simeonidis and Mitrakas 2021). They show promise for the extraction of a variety of metals, including transition and noble metals, especially those based on ammonium and phosphonium cations. Functionalized ILs improve extractability and selectivity, including the effective extraction of metals like Cu and Cd using thiosalicylate derivatives. Even though metal recovery presents certain difficulties, techniques like back extraction and the preparation of task-specific ILs are essential for increasing the sustainability and efficiency of metal extraction from wastewater (Yudaev and Chistyakov 2022).

The pyrrolidinium-based ILs, namely *N*-(2-hydroxyethyl)-*N*-methylpyrrolidinium bis(trifluoromethylsulfonyl)imide ([Pyr-EA][NTf_2]), have shown outstanding activity in extracting gold from e-waste, with a high capacity of 524.6 mg/g. The presence of an ester group as an N-substituent increases its extraction efficacy, highlighting the necessity of task-specific IL design. Although the study only looks at pyrrolidinium ILs and does not compare imidazolium or phosphonium types of ILs, it demonstrates the effectiveness and reusability of [Pyr-EA][NTf_2] in sustainable and selective gold recovery (Liu et al. 2023a, b). Dispersive liquid–liquid microextraction is used to extract zinc from environmental water samples utilizing [BMIM][PF_6]. By dispersing [BMIM][PF_6] in a solvent to form tiny droplets, this method enhances interaction with Zn ions for efficient extraction. With its high sensitivity, rapid extraction, and lack of harmful organic solvents, the technique is an effective and environmentally friendly way to test trace Zn in water (A, R and A 2021).

3.7 Conclusion

The versatility of ILs and their growing use in a range of scientific and industrial fields are thoroughly illustrated in this chapter. By enhancing medication solubility, stability, permeability, and bioavailability, they offer the pharmaceutical industry innovative methods for therapeutic formulation and targeted delivery. In order to create high-performance biosensors and electrochemical detection systems, ILs are essential in diagnostics as well since they improve signal sensitivity, stability, and biocompatibility. It is mentioned that

ILs are used in the extraction of bioactive compounds to increase their efficiency and selectivity, as well as how they behave in MAE, UAE, and SLE. Their capacity to form reusable, task-specific systems is consistent with green chemistry concepts, encouraging environmentally beneficial behavior and reducing hazardous waste. Future developments in green chemistry, biomedical science, and environmental protection are anticipated to be significantly facilitated by ILs as research into their design and utility continues.

References

A, P., R, V. and A, B.L. (2021) 'Dispersive liquid-liquid microextraction of zinc from environmental water samples using ionic liquid', *Chemical Engineering Communications*, 208(6), pp. 914–923. Available at: https://doi.org/10.1080/00986445.2020.1865935

Abbas, Y. *et al.* (2020) 'Heteroatom-Doped Carbon Nanoparticle–Ionic Liquid Composites as Electrochemical Sensors for Uric Acid', *ACS Applied Nano Materials*, 3(11), pp. 11383–11390. Available at: https://doi.org/10.1021/acsanm.0c02466

Akbarzadeh Gondoghdi, P. *et al.* (2024) 'Effect of 2-hydroxyethylammonium carboxylate protic ionic liquids on the solubility and cytotoxicity of indomethacin', *BMC Chemistry*, 18(1), p. 109. Available at: https://doi.org/10.1186/s13065-024-01212-4

Albasri, S.A.M.M. and Wahab, H.S. (2020) 'Removal of Acetaminophen Residues from Wastewater by Bulk Liquid Membrane Process', *Iraqi Journal of Chemical and Petroleum Engineering*, 21(4), pp. 1–9. Available at: https://doi.org/10.31699/IJCPE.2020.4.1

Atta, A.M. *et al.* (2025) 'Application of new tetra-cationic imidazolium ionic liquids for capture and conversion of CO2 to amphiphilic calcium carbonate nanoparticles as a green additive in water based drilling fluids', *Journal of Environmental Sciences*, 150, pp. 159–176. Available at: https://doi.org/10.1016/j.jes.2024.02.028

Bagheri, M. *et al.* (2024) 'Effect of novel surfactant protic ethanolamine based ionic liquids on the aqueous solubility and thermodynamic properties of acetaminophen', *Journal of Molecular Liquids*, 406, p. 124999. Available at: https://doi.org/10.1016/j.molliq.2024.124999

Bekdemir, A. *et al.* (2022) 'Ionic Liquid-Mediated Transdermal Delivery of Thrombosis-Detecting Nanosensors', *Advanced Healthcare Materials*, 11(11). Available at: https://doi.org/10.1002/adhm.202102685

Bhowmik, D. *et al.* (2024) 'An efficient, green and micellar catalyzed preparative-scale synthesis of 3,4-Disubstituted isoxazole-5(4H)-ones in the water', *Sustainable Chemistry for the Environment*, 5, p. 100070. Available at: https://doi.org/10.1016/j.scenv.2024.100070

Chaudhary, A. and Srivastava, R. (2025) 'Bridged N–based ionic liquid and their application in synthesis of five membered N–heterocycles', *Tetrahedron*, 183, p. 134721. Available at: https://doi.org/10.1016/j.tet.2025.134721

Dr. D. Valli Sowbhagyam (2024) 'Ionic Liquids as Green Solvents: A Comprehensive Review', *International Research Journal on Advanced Engineering Hub (IRJAEH)*, 2(02), pp. 220–224. Available at: https://doi.org/10.47392/IRJAEH.2024.0035

Ehsan Kianfar and Sajjad Mafi (2020) 'Ionic Liquids: Properties, Application, and Synthesis', *Fine Chemical Engineering*, pp. 22–31. Available at: https://doi.org/10.37256/fce.212021693

Feng, J. *et al.* (2020) 'Recent advances of ionic liquids in sample preparation', *TrAC Trends in Analytical Chemistry*, 125, p. 115833. Available at: https://doi.org/10.1016/j.trac.2020.115833

Ferreira, B.L., Granato, D. and Nunes, I.L. (2023) 'Uses of ionic liquids to obtain bioactive compounds: insights from the main international regulations for technological applications', *Critical*

Reviews in Food Science and Nutrition, 63(28), pp. 9217–9232. Available at: https://doi.org/10.1080/10408398.2022.2067115

Flora, F.C. *et al.* (2023) 'Combined Use of Ionic Liquid-Based Aqueous Biphasic Systems and Microfluidic Devices for the Detection of Prostate-Specific Antigen', *Biosensors*, 13(3), p. 334. Available at: https://doi.org/10.3390/bios13030334

Franco-Vega, A. *et al.* (2021) 'Effect of imidazolium ionic liquids as microwave absorption media for the intensification of microwave-assisted extraction of Citrus sinensis peel essential oils', *Chemical Engineering and Processing - Process Intensification*, 160, p. 108277. Available at: https://doi.org/10.1016/j.cep.2020.108277

Fu, S. *et al.* (2024) 'An antistatic agent based on polyionic liquid applied to nylon 6', *Green Chemical Engineering*, 5(3), pp. 399–407. Available at: https://doi.org/10.1016/j.gce.2023.10.001

Gaikwad, N., Kudal, S. and Avachat, A.M. (2022) 'Choline-Amino Acid-Derived Bio-ionic Liquids for Solubility Enhancement of Zafirlukast', *AAPS PharmSciTech*, 23(5), p. 146. Available at: https://doi.org/10.1208/s12249-022-02296-8

Ghodse, S.M. *et al.* (2018) 'Dual Utility of Heterogeneous Catalyst ZSM-5 for C–C Cleavage Leading to Nitriles, and for the Synthesis of Hydrazides', *ChemistrySelect*, 3(16), pp. 4168–4172. Available at: https://doi.org/10.1002/slct.201800690

Ghorbanizamani, F., Moulahoum, H. and Timur, S. (2024) 'Ionic liquid-reinforced Hydroxyapatite@nano-TiO2 as a green platform for Immuno-electrochemical sensing applications', *Talanta*, 280, p. 126688. Available at: https://doi.org/10.1016/j.talanta.2024.126688

Goutham, R. *et al.* (2022) 'Ionic liquids in wastewater treatment: A review on pollutant removal and degradation, recovery of ionic liquids, economics and future perspectives', *Journal of Molecular Liquids*, 349, p. 118150. Available at: https://doi.org/10.1016/j.molliq.2021.118150

Gulhane, A.S. *et al.* (2025) 'Optimizing Solvent Systems for Effective Lutein Solvation: A Molecular Dynamics Approach Using 2-Methyl Tetrahydrofuran, Deep Eutectic Solvent, Microemulsion, and Hexane', *Macromolecular Theory and Simulations* [Preprint]. Available at: https://doi.org/10.1002/mats.202500016

Al Hassan, M.K. *et al.* (2023) 'Ionic liquids and NADES for removal of organic pollutants and heavy metals in wastewater: A comprehensive review', *Journal of Molecular Liquids*, 391, p. 123163. Available at: https://doi.org/10.1016/j.molliq.2023.123163

Hassanpour, M. *et al.* (2022) 'Ionic liquid-mediated synthesis of metal nanostructures: Potential application in cancer diagnosis and therapy', *Journal of Ionic Liquids*, 2(2), p. 100033. Available at: https://doi.org/10.1016/j.jil.2022.100033

Hatvate, N. and Gupta, J. (2024) 'An efficient, green and solvent-free three-component synthesis of 2,4,6-triarylpyridines utilizing ZSM-5 as a heterogeneous catalyst', *Synthetic Communications*, 54(20), pp. 1760–1770. Available at: https://doi.org/10.1080/00397911.2024.2403779

Hatvate, N.T. *et al.* (2018a) 'Transition metal free large-scale synthesis of aromatic vinyl chlorides from aromatic vinyl carboxylic acids using bleach', *Tetrahedron Letters*, 59(43), pp. 3892–3894. Available at: https://doi.org/10.1016/j.tetlet.2018.09.034

Hatvate, N.T., Ghodse, S.M. and Telvekar, V.N. (2018b) 'Metal-free synthesis of 2-aminothiadiazoles via TBHP-Mediated oxidative C-S bond formation', *Synthetic Communications*, 48(3), pp. 285–290. Available at: https://doi.org/10.1080/00397911.2017.1398330

Hatvate, N.T. *et al.* (2020) 'Metal-Free Synthesis of Pyrimidinone Derivatives via Biginelli Reaction Using Aqueous NaICl2', *Letters in Organic Chemistry*, 17(8), pp. 613–617. Available at: https://doi.org/10.2174/1570178617666191126095808

Hatvate, N.T. and Ghodse, S.M. (2020) 'One-pot three-component synthesis of isoxazole using ZSM-5 as a heterogeneous catalyst', *Synthetic Communications*, 50(23), pp. 3676–3683. Available at: https://doi.org/10.1080/00397911.2020.1815786

Hatvate, N.T., Ghodse, S.M. and Telvekar, V.N. (2019) 'ZSM-5 catalyzed one pot three-component synthesis of 5-substituted-1H-tetrazoles from aldehyde', *Synthetic Communications*, 49(24), pp. 3553–3559. Available at: https://doi.org/10.1080/00397911.2019.1679185

Hayyan, A. *et al.* (2025) 'Ultrasonic assisted extraction of oil from argan seeds using ionic liquids as novel co-solvent', *Biomass Conversion and Biorefinery*, 15(6), pp. 9463–9473. Available at: https://doi.org/10.1007/s13399-024-05847-0

Hu, Y. *et al.* (2023) 'Ionic liquids revolutionizing biomedicine: recent advances and emerging opportunities', *Chemical Society Reviews*, 52(20), pp. 7262–7293. Available at: https://doi.org/10.1039/D3CS00510K

Islam, Md.R. *et al.* (2021) 'Insulin Transdermal Delivery System for Diabetes Treatment Using a Biocompatible Ionic Liquid-Based Microemulsion', *ACS Applied Materials & Interfaces*, 13(36), pp. 42461–42472. Available at: https://doi.org/10.1021/acsami.1c11533

Isosaari, P., Srivastava, V. and Sillanpää, M. (2019) 'Ionic liquid-based water treatment technologies for organic pollutants: Current status and future prospects of ionic liquid mediated technologies', *Science of The Total Environment*, 690, pp. 604–619. Available at: https://doi.org/10.1016/j.scitotenv.2019.06.421

Jiang, B. *et al.* (2023) 'A simple hydroxypyridine ionic liquids for conversion of CO2 into quinazoline-2,4(1H,3H)-diones under atmospheric conditions', *Chinese Journal of Chemical Engineering*, 62, pp. 217–224. Available at: https://doi.org/10.1016/j.cjche.2023.04.003

Khoo, K.S. *et al.* (2021) 'Permeabilization of Haematococcus pluvialis and solid-liquid extraction of astaxanthin by CO2-based alkyl carbamate ionic liquids', *Chemical Engineering Journal*, 411, p. 128510. Available at: https://doi.org/10.1016/j.cej.2021.128510

Khraisheh, M. *et al.* (2021) 'Ionic liquids application for wastewater treatment and biofuel production: A mini review', *Journal of Molecular Liquids*, 337, p. 116421. Available at: https://doi.org/10.1016/j.molliq.2021.116421

Kosiński, S. *et al.* (2022) 'Ionic liquids as antistatic additives for polymer composites – A review', *Polymer Testing*, 112, p. 107649. Available at: https://doi.org/10.1016/j.polymertesting.2022.107649

Kuddushi, M. *et al.* (2024) 'Review of ionic liquid and ionogel-based biomaterials for advanced drug delivery', *Advances in Colloid and Interface Science*, 331, p. 103244. Available at: https://doi.org/10.1016/j.cis.2024.103244

Li, L. *et al.* (2023) 'Ionic Liquids: Momentous Tools in Transdermal Delivery of Biomacromolecules', *Advanced Therapeutics*, 6(6). Available at: https://doi.org/10.1002/adtp.202200332

Li, X. *et al.* (2022) 'Applications of choline-based ionic liquids in drug delivery', *International Journal of Pharmaceutics*, 612, p. 121366. Available at: https://doi.org/10.1016/j.ijpharm.2021.121366

Lim, J.R., Chua, L.S. and Mustaffa, A.A. (2022) 'Ionic liquids as green solvent and their applications in bioactive compounds extraction from plants', *Process Biochemistry*, 122, pp. 292–306. Available at: https://doi.org/10.1016/j.procbio.2022.10.024

Liu, P. *et al.* (2023) 'Molecular Insights into the Binding Behavior of Imidazolium Ionic Liquids to the Receptor Binding Domain of the SARS-CoV-2 Spike Protein', *The Journal of Physical Chemistry B*, 127(20), pp. 4396–4405. Available at: https://doi.org/10.1021/acs.jpcb.3c00593

Liu, X. *et al.* (2023) 'Selective and Efficient Gold Extraction from E-Waste by Pyrrolidinium-Based Ionic Liquids with Various N-Substituents', *ACS Sustainable Chemistry & Engineering*, 11(2), pp. 638–648. Available at: https://doi.org/10.1021/acssuschemeng.2c05455

Lu, B. *et al.* (2022) 'Ionic liquid transdermal delivery system: Progress, prospects, and challenges', *Journal of Molecular Liquids*, 351, p. 118643. Available at: https://doi.org/10.1016/j.molliq.2022.118643

Moshikur, R.M. *et al.* (2021) 'Favipiravir-Based Ionic Liquids as Potent Antiviral Drugs for Oral Delivery: Synthesis, Solubility, and Pharmacokinetic Evaluation', *Molecular Pharmaceutics*, 18(8), pp. 3108–3115. Available at: https://doi.org/10.1021/acs.molpharmaceut.1c00324

Neamah, A. jaber, Ghareeb, M.M. and Rashid, A.M. (2022) 'Ionic liquid: A green designer solvent for topical delivery of Nano system', *Maaen Journal for Medical Sciences*, 1(1). Available at: https://doi.org/10.55810/2789-9128.1008

Niu, J. *et al.* (2024) 'Acetate-based ionic liquid immobilized Fe-MIL-101-NH2: A highly efficient heterogeneous catalyst for the conversion of CO2 into oxazolidinones with N-aryl epoxy amines', *Journal of Environmental Chemical Engineering*, 12(5), p. 113503. Available at: https://doi.org/10.1016/j.jece.2024.113503

Padinhattath, S.P., Shaibuna, M. and Gardas, R.L. (2025) 'Exploring Ionic Liquid-Based Liquid–Liquid Extraction as a Benign Alternative for Sustainable Wastewater Treatment', *Sustainability & Circularity NOW*, 02(CP). Available at: https://doi.org/10.1055/a-2515-6076

Parit, S. *et al.* (2024) 'Comparative review on homogeneous and heterogeneous catalyzed synthesis of 1,3-thiazole', *Synthetic Communications*, pp. 1–21. Available at: https://doi.org/10.1080/00397911.2024.2399187

Patil, B.R. *et al.* (2023) 'Ultrasound-assisted facile and efficient synthesis of novel Benzoxazole derivatives from o-aminocardanol using Indion 190 resin as a reusable catalyst', *Journal of Chemical Sciences*, 135(1), p. 4. Available at: https://doi.org/10.1007/s12039-022-02120-7

Pedro, S.N. *et al.* (2021a) 'Ionic Liquids in Drug Delivery', *Encyclopedia*, 1(2), pp. 324–339. Available at: https://doi.org/10.3390/encyclopedia1020027

Pedro, S.N. *et al.* (2021b) 'Ionic Liquids in Drug Delivery', *Encyclopedia*, 1(2), pp. 324–339. Available at: https://doi.org/10.3390/encyclopedia1020027

Przypis, M. *et al.* (2020) 'Inexpensive and tuneable protic ionic liquids based on sulfuric acid for the biphasic synthesis of alkyl levulinates', *Journal of Molecular Liquids*, 308, p. 113166. Available at: https://doi.org/10.1016/j.molliq.2020.113166

Rawat, S., Kumar, A. and Bhaskar, T. (2022) 'Ionic liquids for separation of lignin and transformation into value-added chemicals', *Current Opinion in Green and Sustainable Chemistry*, 34, p. 100582. Available at: https://doi.org/10.1016/j.cogsc.2021.100582

Ray, A. and Saruhan, B. (2021) 'Application of Ionic Liquids for Batteries and Supercapacitors', *Materials*, 14(11), p. 2942. Available at: https://doi.org/10.3390/ma14112942

Reddy, R.R. *et al.* (2018) 'Selective binding and dynamics of imidazole alkyl sulfate ionic liquids with human serum albumin and collagen – a detailed NMR investigation', *Physical Chemistry Chemical Physics*, 20(14), pp. 9256–9268. Available at: https://doi.org/10.1039/C7CP08298C

Roque-Diaz, Y. *et al.* (2021) 'The Dark Side of Platinum Based Cytostatic Drugs: From Detection to Removal', *Processes*, 9(11), p. 1873. Available at: https://doi.org/10.3390/pr9111873

Shukla, M.K. *et al.* (2023) 'Role and Recent Advancements of Ionic Liquids in Drug Delivery Systems', *Pharmaceutics*, 15(2), p. 702. Available at: https://doi.org/10.3390/pharmaceutics15020702

Simeonidis, K. and Mitrakas, M. (2021) 'Technologies Developing in Heavy Metals' Removal from Water', *Water*, 13(6), p. 860. Available at: https://doi.org/10.3390/w13060860

Singh, P.P., Kanodia, S. and Ambika (2024) 'Ionic Liquids as a Non-conventional Catalyst for the Synthesis of N -heterocycles', *Mini-Reviews in Organic Chemistry*, 21(7), pp. 742–763. Available at: https://doi.org/10.2174/1570193X20666230507183754

Wang, J. *et al.* (2024) 'Application of ionic liquid ultrasound-assisted extraction (IL-UAE) of lycopene from guava (Psidium guajava L.) by response surface methodology and artificial neural network-genetic algorithm', *Ultrasonics Sonochemistry*, 106, p. 106877. Available at: https://doi.org/10.1016/j.ultsonch.2024.106877

Wu, X. *et al.* (2019) 'Improving dermal delivery of hydrophilic macromolecules by biocompatible ionic liquid based on choline and malic acid', *International Journal of Pharmaceutics*, 558, pp. 380–387. Available at: https://doi.org/10.1016/j.ijpharm.2019.01.021

Yuan, J. *et al.* (2022) 'Ionic liquids as effective additives to enhance the solubility and permeation for puerarin and ferulic acid', *RSC Advances*, 12(6), pp. 3416–3422. Available at: https://doi.org/10.1039/D1RA07080K

Yudaev, P.A. and Chistyakov, E.M. (2022) 'Ionic Liquids as Components of Systems for Metal Extraction', *ChemEngineering*, 6(1), p. 6. Available at: https://doi.org/10.3390/chemengineering6010006

Zhu, T. *et al.* (2022) 'Ionic liquid and ultrasound-assisted extraction of chestnut shell pigment with good hair dyeing capability', *Journal of Cleaner Production*, 335, p. 130195. Available at: https://doi.org/10.1016/j.jclepro.2021.130195

Zhuo, Y. *et al.* (2024) 'Ionic Liquids in Pharmaceutical and Biomedical Applications: A Review', *Pharmaceutics*, 16(1), p. 151. Available at: https://doi.org/10.3390/pharmaceutics16010151

Emerging Trends and Opportunities 4

Abstract

Ionic liquids (ILs) have emerged as a versatile class of compounds that offer unique physicochemical properties, transforming the field of sustainable chemistry. This chapter explores the role of ILs in nano-catalysis, highlighting their function as green solvents and stabilizing agents in catalytic reactions. Their applications in sustainable energy, such as fuel cells, batteries, and CO_2 capture, underscore their potential to address global energy challenges. The development and use of biodegradable ionic liquids further enhance their environmental compatibility, paving the way for eco-friendly industrial applications. Finally, the chapter discusses emerging trends and future directions in IL research, including task-specific ILs, deep eutectic solvents, and hybrid systems, providing insights into the next frontier of green and sustainable chemistry.

4.1 Introduction

The last several decades have brought a remarkable rise in research focus on ionic liquids (ILs) because of their exclusive properties, including low volatility combined with high thermal stability, high absorption, negligible volatility, and customizable features (Valli Sowbhagyam 2024). ILs based exclusively on ions provide innovative approaches toward solving problems in multiple industries, including energy production, environmental sustainability, and catalytic processes. The ongoing research on ILs creates new trends that transform numerous scientific areas and technological sectors. Through nano-catalysis employing ionic liquids, researchers progress in both scientific understanding and industrial applications by developing sustainable energy applications and biodegradable ionic liquids. This opens the doors for a wide range of applications in various fields for these ILs (Zhou et al. 2023; Islam et al. 2023).

H. N. Akolkar et al., *Ionic Liquids*, Synthesis Lectures on Sustainable Development,
https://doi.org/10.1007/978-3-032-02746-7_4

4.2 Ionic Liquid-Based Nano-Catalysis

The concept of green chemistry encourages sustainable methodology development, which prevents toxic agents from entering the catalyst synthesis process (Hatvate and Ghodse 2020; Hatvate et al. 2018, 2019, 2020, 2024; Bhowmik et al. 2024; Gadkari et al. 2020, 2021a, b; Telvekar et al. 2020; Hatvate and Ghodse 2020). Nanotechnology, especially nanoparticles, plays a vital role in green growth through applications in drug delivery, pollution control, catalysis, and more (Gholap et al. 2024a, b, c, 2025; Pardeshi et al. 2024). The promise of metal NPs encounters two significant barriers: agglomeration and low stability. Adding ligands and polymeric materials to metal nanoparticles as stabilizing agents improves their durability (Bhowmik et al. 2025). Ionic liquids serve as effective dispersing vehicles for metal nanoparticles because they improve both catalyst conversion rates and solution particle distribution quality. Adopting supported IL systems helps solve industrial constraints such as high costs and mass transfer problems encountered in IL operations. IL-based nanomaterials have a variety of applications, such as biocatalysis, biodiesel production, and hydrogen generation. For biocatalysis, supported IL phases and supported IL-like phases are utilized for immobilization on nanomaterials and enzyme stabilization, increased enzyme efficacy, and reusability (Nasri et al. 2022; Wolny et al. 2021). Ester hydrolysis, triacetin hydrolysis, biodiesel production wastewater treatment, biosensors, and CO_2 separation are areas where these biocatalysts have found their use. Nanomaterials like carbon nanotubes, silica, and magnetic nanoparticles improve enzyme performance (Wolny et al. 2021). For biodiesel production, an alkaline IL nano-catalyst (NC) (Fe_3O_4-SiO_2-based IL) was used to mediate the transesterification of algal oil into biodiesel. The NC showed high surface area, saturation magnetization, and thermal stability, resulting in effective conversion with fantastic reusability. Economically sound, environmentally friendly, with a minimized carbon footprint, biodiesel synthesis can be achieved at an industrial scale using this NC (Sangeetha et al. 2024). Additionally, IL-based NCs (IL-NCs) are utilized for electrocatalytic and photocatalytic H_2 evolution reactions and generation via hydrolysis of hydrides. They can also be used in electrochemical redox reactions like oxygen evolution, hydrogen evolution, and oxygen reduction reactions, as well as photocatalytic reactions like photocatalytic hydrogen evolution and water splitting. A novel interest has also been seen in applying ILs in synthesizing catalysts for energy technologies (Nasri et al. 2022).

4.2.1 ILs in Pollution Control

Since ILs are very viscous, their use as catalysts is difficult as they are challenging to recover. Porous materials and magnetic nanoparticles are now being used to remedy this problem with IL. These solid, magnetically responsive ILs have been successfully utilized for various applications. Two magnetic nanocomposite catalysts were developed for the efficient desulfurization of sulfur-containing compounds such as thiobenzoic acid and 2,2′-Dinitro-5,5′-dithiodibenzoic acid (Aziz et al. 2025). Nitroaromatic compounds are a category of materials that are widely used but are incredibly harmful to the environment as well as human health. A novel toughened IL hydrogel was produced by polymerizing acrylamide and ILs with SiO_2@poly(butyl acrylate) particles that showed high mechanical strength and enhanced stretchability. This hydrogel was later loaded with palladium NPs, and this nano-catalyst efficiently degraded p-nitrophenol with an efficiency of 90% even after 15 uses. This approach was seen as a promising candidate for water treatment (Lv et al. 2024).

In a study, new dihydrobenzopyrazoloazepine derivatives were synthesized using Ag/Fe_3O_4/CdO functionalized multi-wall carbon nanotubes as high-performance catalysts in IL (1-Octhyl-3-methyl imidazolium bromide) as stabilizer and soft template. The nano-catalyst showed successful reduction of 4-nitrophenol in water under mild conditions. The catalysts showed short reaction time and easy recovery (Kadhim et al. 2023). In another study, a green catalyst (PILs-FPS) was produced by integrating tetraethyl orthosilicate with tripolyphosphate on fibrous phosphosilicate structures while incorporating poly (ILs). Using the nano-catalyst, limonene epoxide and CO_2 were transformed into bio-based cyclic carbonates through an effective solvent-free reaction at temperate conditions. The carbonates obtained through this process were processed into sustainable polymeric materials. Obtaining high adsorption capacity along with simple catalyst recovery was possible because of the FPS structure. Multiple reuse cycles were possible for PILs-FPS while maintaining high selectivity and excellent performance without any activity loss. The method presents an environmental method to use CO_2 resources and make polymers (Monavari et al. 2024).

An eco-friendly method to synthesize advanced dandelion-shaped TiO_2 with an anatase/TiO_2 phase junction and high surface area utilizing deep eutectic solvents as a fabrication solution has recently been prepared. The 3D arrangement built from 2D mesoporous nanosheets elevates the catalytic performance. A uniform distribution of polyoxomolybdate\[Mo132] nanospheres with ionic liquid groups took place on dendritic fibrous nano-titanium (DFNT) to produce DFNT\@IL/Mo_{132} composites. The active sites for CO_2 chemisorption, as well as the ability to control fiber sizes, are made possible by using ILs (Chen et al. 2023). In a study, a supramolecular catalyst composed of fluorenyl-modified amino acids together with $Cu2^{+}$ was produced as an effective catalyst to break down native lignin with the help of IL conditions. The catalyst imitates catechol oxidase functions through stabilization mechanisms, including aromatic stacking forces and precise amino acid/halide bonding. The technology enables successful lignin degradation processes that do not require the addition of electron mediators. The catalyst shows excellent thermal stability while operating well in ionic liquids, which makes it a suitable solution for sustainable lignin valorization on an industrial production scale (Liu et al. 2024).

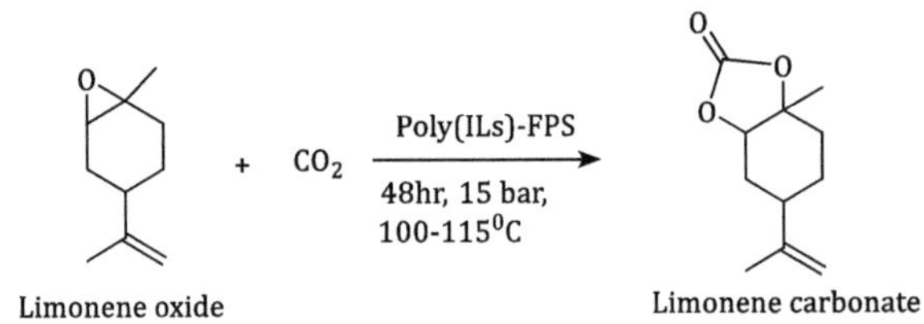

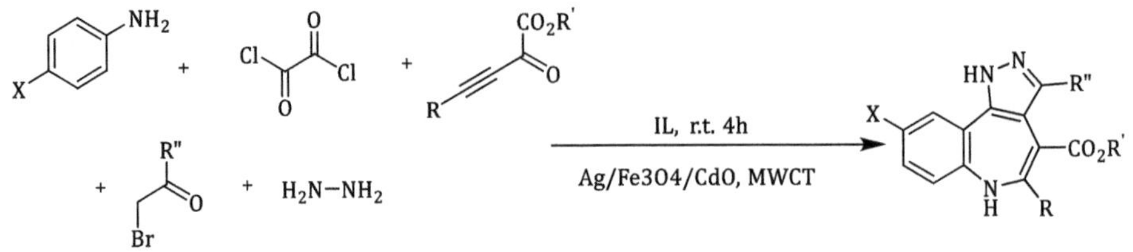

4.2.2 ILs for the Synthesis of Heterocyclic Compounds

ILs display vigorous catalytic activity and excellent stability, making them highly efficient in several chemical processes. Attributed to these properties, ILs are widely utilized as activating, protective, and stabilizing agents (Bakhshali-Dehkordi et al. 2020). ILs are commonly used in organic reactions like Diels-Alder and Friedel-Crafts for their catalytic properties and ability to act as reaction media. Since the first reported use in 1986, they have performed fantastically in various catalytic reactions. A key feature of ILs is their reusability, rendering them environmentally and economically beneficial (Moradi et al. 2024). IL-based titanium oxide nano-catalysts were used for the synthesis of 1*H*-pyrazolo[1,2-b]phthalazine-2-carbonitrile derivatives using phthalhydrazide, malononitrile, and aldehydes, which gave high efficiency (Hamidinasab et al. 2020). In another study, one-pot synthesis of quinazolinone was done by using 3-amino-1,2,4-triazole or 2-aminobenzimidazole, dimedone, and aromatic aldehydes, catalyzed by titanium oxide magnetic nanoparticles functionalized with DABCO-based ILs (Bakhshali-Dehkordi et al. 2020). Methylimidazole-based ILs supported on silica-coated perlite NPs were used to synthesize 1*H*-pyrazolo [1,2-*b*] phthalazine-5,10-diones by using phthalimide, hydrazine hydrate, various aryl aldehydes, and malononitrile (Moradi et al. 2024).

In another study, carbon nanotubes were used as a template for the synthesis of magnetic mesoporous nanotubes, which were then further functionalized using acidic IL and utilized as a heterogeneous catalyst for the synthesis of indeno[1,2-*b*]indole-9,10-dione derivatives (Bistgani et al. 2024). A comparative study was performed by immobilizing piperazine-based ILs on ZnO and SiO_2 NPs to synthesize benzoxazoles and benzimidazoles. Further, ZnO-IL-based catalysts, which showed better performance, showed adequate *N*-Boc protection of amines (Mousapour et al. 2024). In another study, amorphous silica was bonded with dicationic IL (TSPSED) and used as a nano-catalyst for the generation of bis(pyrazolyl)methanes and pyrazolopyranopyrimidines, which gave high yield and short reaction time (Jaddan et al. 2023) (Table 4.1).

4.2.3 ILs in Hydrogenation Reactions

The key role played by ILs in hydrogenation reactions is to immobilize and stabilize the 3d metal NPs that are known for conducting these reactions (Zenner et al. 2025). Additionally, ILs are also used as media for the synthesis of NPs, where the ILs act as both dispersion media and stabilizers (Krishnan et al. 2024). In a study, copper-based magnetic nano-catalysts were synthesized, which performed catalytic transfer hydrogenation reactions on biomass-derived furfural (FF) to generate furfuryl alcohol. The NPs were supported on imidazole-functionalized magnetic materials (imidazolium-derived IL). The XPS analysis revealed that the IL linking with imidazolium produced a positive surface polarization on copper NPs. The catalyst retains its potential as a selectivity-reliable and reusable option for green chemical production, but the activity drops 30% after five cycles of use (Gaete et al. 2025). In another study, the hydrogenation was carried out by designing recyclable catalysts using 1,1,3,3-tetramethyl guanidine (TMG)-based IL hydrogels. Palladium NPs were dispersed into the hydrogel network, facilitated by DSA's hydrophobic long-chain structure, preventing Pd aggregation. This nano-catalyst showed excellent hydrophobicity and 99% efficiency in the hydrogenation of styrene within 30 min. Additionally, this catalyst also demonstrated good reusability (Li et al. 2024).

A study produced bimetallic manganese–ruthenium nanoparticles (Mn_xRu_{100-x}) prepared on imidazolium-based IL phase (SILP). The bimetallic ratio between Mn and Ru directly impacted benzylideneacetone reduction, both in terms of reaction rate and selectivity, although metal alloying did not occur. The activation of aromatic ring hydrogenation changes from full hydrogenation with $Mn_{15}Ru_{85}$ to complete lack of hydrogenation with $Mn_{25}Ru_{75}$. Bimetallic catalyst synergy can appear independently of alloy formation, thus opening new pathways for creating selective hydrogenation catalysts (Zenner et al. 2025).

In another study, new Ni and Ni-NiO NPs were synthesized by opting for an organometallic approach with Ni NPs in two methoxy-functionalized ether-based ILs. Previously reported literature showed that the methoxy group has a stabilizing effect in PEG-based systems. Hence, the ILs were designed to enhance the catalytic performance and dispersion. The resulting nano-catalysts showed promising selectivity when tested for the hydrogenation of 2-cyclohexen-1-one (Krishnan et al. 2024). Another study demonstrated the successful synthesis of bimetallic RuCu nano-catalyst integrated on l-cysteine functionalized IL-modified magnetic NPs. The bimetallic catalyst showed exceptional effectiveness when reducing various forms of nitrophenol isomers (ortho, meta, and para)

Table 4.1 Summarized synthesis of heterocycles

Sr No	Catalysts	Solvent	Conditions (°C)	Time	Synthesized compound	Yield (%)
1	$NiFe_2O_4\_TiO_2$-IL	PEG 400	90	10 min	1H-pyrazolo[1,2-b]phthalazine-2-carbonitrile derivatives	97
2	$Fe_3O_4\_TiO_2$-IL	Solvent free	80	40 min	Quinazolinone	97
3	Perlite NPs/SiO_2/IL	Ethanol	78	4 h	Pyrazolo phthalazine derivatives	88
4	Magnetic mesoporous silica nanotubes	Water	75	70 min	indeno[1,2-*b*]indole-9,10-dione derivatives	96
5	ZnO-IL	Solvent free	60	5 min	*N*-Boc protected aniline	95
6	Nano-(TSPSED)(Cl_2)	Solvent free	50	6 min	bis(pyrazolyl)methanes	96
7			60	10min	pyrazolopyranopyrimidines	97

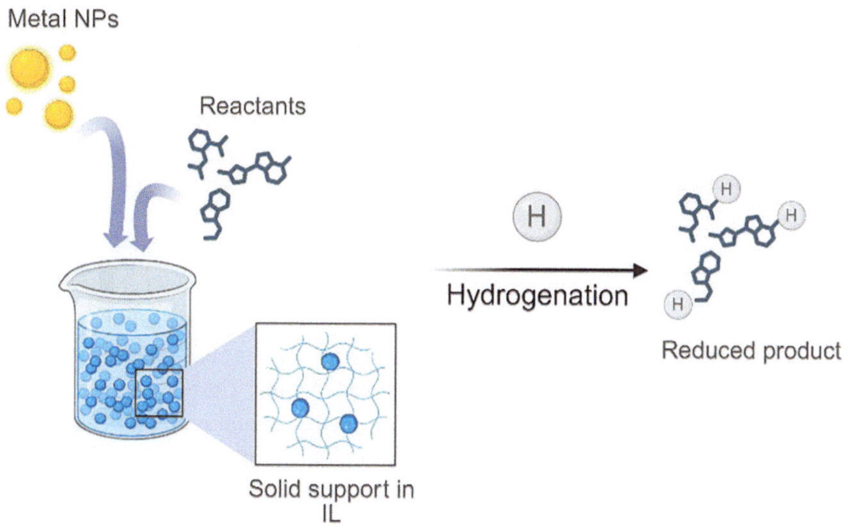

Fig. 4.1 Hydrogenation reaction by ILs

within 9–16 min while degrading organic dyes (4–7 min) and decomposing tetracycline (30 min) contaminants in water under temperature conditions. The combination of Ru-Cu catalyst demonstrated concentrated performance compared to single-metal systems, producing enhanced outcomes by factors ranging from two to thirty-three in different reactions. The efficient catalyst recycling occurred 14 times, without activity reduction being detected (Gholinejad et al. 2024) (Fig. 4.1).

4.2.4 ILs in Coupling Reactions

ILs significantly impact the performance of Ullmann coupling, attributed to their highly polar environment, by interacting strongly with the Cu NPs surface, which increases the stability and ensures even distribution of the catalyst particles. Although ILs provide various benefits, they are limited in nature as they can cause agglomeration. To overcome this obstacle, supported IL phases have been introduced as a powerful alternative to the existing ILs. A study showed the impact of replacing anion species within ILs-functionalized silica supports on catalyst productivity for CuNP-based catalysts during C–N Ullmann coupling reactions. Iodide anion (I^-) resulted in 100% conversion, but BF_4^- anion gave the highest selectivity rate at 57%. Computational analyses demonstrated that substrate bonding occurs via coordinative covalent bonds with polarized interactions, while aniline paths achieve maximum stability. The selected anion affected substrate stabilization and reaction kinetics, alongside spin density, which directly influenced catalytic effectiveness and yield control. The composition of ionic liquids proved essential for creating nanocatalysts, which led to optimal performance in sustainable organic synthesis (Gaete et al. 2024).

4.3 Sustainable Energy Applications

The sustenance of energy supply has emerged as one of the crucial challenges of the current times, further enforced by the contribution of fossil fuels in climate change, exploited reserves, air quality concerns, and energy security. The increased utilization of renewable sources such as solar and wind is reassuring, but their recurrent use requires the development of large-scale energy storage and variable load solutions. Carbon capture technology has been developed in parallel with it to mitigate environmental pollution caused by fossil fuel usage. Attributing to these factors, the development of materials having enhanced energy generation, storage, and delivery capabilities is pivotal. ILs are composed of ions and act as liquid salts. They have attracted attention due to their low evaporation rate, steady chemical and electrochemical characteristics, and adjustable features. Attributing to their good ionic conduction, they are helpful in advanced batteries, dye-sensitized solar cells, fuel cells, double-layer capacitors, thermo-cells, actuators, and systems for splitting water. Apart from electrochemical uses, ILs are being studied as heating and cooling liquids, materials for energy storage, and as part of processes involving CO_2, CO_2 capture, treated biomass, biofuel extraction, and as fuel propellant in aerospace. Although ILs can adapt to many different uses, scientists are still struggling to discover the most suitable compounds for designated purposes (Zhou et al. 2023; MacFarlane et al. 2014; Choudhary et al. 2023).

4.3.1 ILs for Energy Storage

IL-based gels can be utilized as safe electrolytes in lithium batteries, aluminium-ion batteries, and supercapacitors as opposed to their counterpart solvents, which are high risk due to their inflammability. The main contributing factor for the performance of ILs in energy storage is their excellent ionic conductivity and expansive electrochemical windows. Many imidazole-based ILs and pyrrolium-based ILs have been reported as suitable materials for superconductors (SCs) and Lithium batteries. The addition of pyrrolidine cation has been shown to improve the ionic conductivity of the IL-based electrolytes significantly. Despite its many perks, ion separation and increased viscosity at lower temperatures compromise the charge transport and ultimately limit the capacitance. These issues present challenges in energy storage systems for the automotive and military sectors, which work in extremely low temperatures (below—60 °C) (Ray et al. 2021).

Eutectic ILs have been developed recently to overcome these challenges, such as a 1-propyl-3-methylpyrrolidinium-based mixture that operates between—70 to 80 °C with a stable 3.5 V window (Karuppasamy et al. 2020a, b). In addition, ionogel electrolytes made from poly (ILs) and polymers are explored for their high voltage capability, self-healing properties, and safety. However, their mechanical stability and self-healing regulation are yet lacking (Karuppasamy et al. 2020a, b).

Attributing to their flexibility and ability to heal themselves, ILs could be part of the emerging materials for rescue gear and endoscopic devices. A study developed a photoresponsive polymer by mixing ILs based on imidazole with polypyrrole, allowing it to act as a material that changes light into heat and conducts electricity. This material worked successfully as a light detector. The study is ongoing on how ILs may be applied in energy systems for conversion and storage (Gao et al. 2021). In a recent study, a novel nanocomposite was prepared using imidazole-based ILs to overcome the supercooling property of phase change materials (PCM) (releases heat by phase transition, i.e., liquid to solid). This nanocomposite showed enhanced thermal conductivity by 36% (Mim et al. 2025). Similarly, in another study, PCM based on IL (tris(2-hydroxyethylammonium)formate) and fatty acids was used, which could show its activity at the range of 50–150 °C and relatively cheaper capacitance (Faraji et al. 2023).

In another study, garnet composites (LALZO) were prepared by dispersing various ILs in small quantities, which showed ~100 times enhanced ionic conductivity compared to the conventional garnet type. This advancement was suspected of the weak, non-uniform chemisorption causing variation at the IL-garnet interface (Kaur et al. 2024). A study developed a new electrolyte using monoethanolamine-based ILs containing anions of formate, acetate, and propionate for activated carbon-based supercapacitors, which showed capacitance retention of 96, 91 and 87%, respectively (Mirzaei-Saatlo et al. 2024). Recently, ILs and NPs hybrid systems known as ionanofluid show good thermophysical properties such as heat storage and transfer (Singh et al. 2022a, b). Using supported liquid membranes (SLMs) can improve how molecules and ions travel in polyelectrolyte membranes (PEMs), which are essential for batteries and fuel cells. Using ionic liquids (ILs) and liquid polymerized ionic liquids (LPILs) in PEMs leads to improvements in molecular and ionic transport. A study demonstrates that by alkylation with 1-methylimidazole, the process can change poly(bromoundecyl acrylate) into an LPIL copolymer. This way uses a standard that links LPILs and SLMs, boosting PEM performance (Li et al. 2023). A similar study synthesized a high-conducting polymer-based electrolyte by doping ILs in a polyethylene oxide complexed sodium iodide matrix. This electrolyte was further used to develop electric double-layer capacitors and Dye-Sensitized Solar Cells, demonstrating high conductivity (Singh et al. 2022a, b).

4.3.2 ILs for Energy Conversion

Hydrogen can be produced from water splitting, with ILs, using solar and renewable energy. ILs have been investigated for use as electrolytes in the water splitting processes and as a medium for synthesizing electrocatalysts that primarily aid in water oxidation. Interestingly, catalysts prepared from ILs have exceeded expectations and performed exceptionally when partnered with IL electrolytes in water splitting (MacFarlane et al. 2014). Efficient electrocatalysts play a key role in advancing sustainable energy technology related to hydrogen and oxygen evolution. In a study, nickel phosphide was synthesized in a hierarchical nano form on the surface of carbon cloth using ILs. Changing the IL content resulted in different $Ni(OH)_2$ precursors and final Ni_2P product morphologies. The nanostructure formed by the ILs on Ni_2P gave the metal very low overpotentials for both HER, with just 132 mV, and OER, with 309 mV, due to its IL-influenced nanostructure offering abundant active sites (Jia et al. 2024). A study focusing on efficient and cost-effective bifunctional electrocatalysts targets water splitting as a vital step in making hydrogen energy. Fe, Cu, or both are used to decorate carbon, using ILs containing the two metals, by carbonization. The presence of Fe_3O_4 and Cu_2O was revealed through characterization. It was found that the Fe, Cu-adorned catalyst had the fastest kinetics in alkaline media with very low onset potential and overpotentials of 325 and 364 mV at 10 and 50 mA cm^2, respectively (Georgijević et al. 2024).

4.3.3 ILs in Fuel Generation from Biomass

Biomass has emerged as an alternative to the ever-depleting non-renewable fossil fuels. This has brought up the need to develop greener options for converting biomass into value-added products like chemicals and fuels. Supported ILs have emerged as superior candidates as they do not possess the drawbacks of conventional ILs. They offer high catalytic activity, tunable properties, and thermal stability, connecting homogeneous and heterogeneous catalysis. ILs were used to convert lignocellulosic biomass by selective deoxygenation into valuable products such as biofuel precursors, 5-hydroxymethyl furfural, and reducing sugar. An additional advantage of the supported ILs is that they are recovered as dry solids, leading to efficient removal by simple decantation (Bohre et al. 2022). A study aimed to improve the lignocellulosic biorefineries by using ILs with lipases together in series to produce valuable goods from lignocellulosic biomass. Additionally, biomass valorization was done by lipase-catalyzed biotransformation in ILs in micro and batch reactors (Šibalić et al. 2023). Hence, ILs provide a green and sustainable alternative to biomass valorization for the generation of biofuels.

4.3.4 ILs for CO_2 Capture and Storage

A basis for reducing climate change is CCS, which absorbs CO_2 emissions from primary point sources like power stations and industrial sites and stores them underground to keep them from escaping into the atmosphere. To alleviate the effects of climate change, CCS entails absorbing high-purity carbon dioxide emissions from sources such as industrial processes, then moving and storing it in geological formations to prevent its release into the atmosphere (Bolarinwa Solanke et al. 2024). A CO_2 capture evaluation of the amino-functionalized IL [TETAH][3-Br-PhO]-PEG200 reveals optimal absorption of 2.37 mol/kg at 298.15 K with a 3:7 mass ratio. The absorbent develops a solid phase and maintains a 72% regeneration efficiency after five cycles at 393.15 K, which is enhanced by interactions with PEG200. With better absorption and recyclability than monoethanolamine (MEA), [TETAH][3-Br-PhO]-PEG200 is positioned as a viable, sustainable non-aqueous CO_2 absorbent (Zhao et al. 2024). The capture and valorization of CO_2 using ILs is a topic of considerable scientific interest. ILs as electrolytes and copper as a cathode would be a more affordable and environment-friendly way to make CO_2 reactive in electrochemical reactions. ILs are used to study CO_2 capture and CO_2 electrochemical reduction since they are greener solvents with special qualities (An et al. 2023).

4.4 Biodegradable ILs

Despite various benefits, the major drawback of ILs is that they are not fully biodegradable, causing toxicity to the environment and organisms. Imidazolium-based ILs are most commonly used but are notoriously resistant to biodegradation and have the potential for bioaccumulation, causing serious environmental harm. These concerns have increased the demand for developing novel, biocompatible, and biodegradable ILs that are not toxic to the environment and organisms. Commonly known, biodegradable ILs are made from naturally occurring chemicals like cholinium cation and amino acid anions, also known as bionic liquids (BILs). Another less frequently used biodegradable IL is pyridium-based ILs (Piedade et al. 2022).

4.4.1 BILs for Biomass Conversion

BILs generate various value-added products from biomass such as cellulose fibres, 5-(hydroxymethyl)furfural, and glucose. The glucose produced can be converted into mycoprotein (artificial meat) or fructose, which is used to compose artificial sweeteners or humectants by isomerization reactions. These increase the scope of BILs in the food, beverage, and pharmaceutical industries to produce protein-containing food from non-food-grade biomass. As mentioned above, ILs can convert biomass into biofuel or jet fuel through cost- and energy-efficient processes. Lignocellulosic biomass is the primary substance used for biofuel production, but other alternatives are also present, such as fermented sugars obtained from sorghum. The BILs are used as a pretreatment for the biomass. The efficiency can be enhanced even more by utilizing sophisticated techniques such as microwaves, ultrasounds, milling, pyrolysis, or steam explosion.

Besides the lignocellulosic biomass, microalgae biomass, Chlorella vulgaris, Spirulina platensis, and chitin were also studied. Microalgae biomass degradation had the drawback of long drying time due to using aqueous systems for pretreatment, and adding other solvents resulted in undesirable interactions. BILs were used to overcome this obstacle as they could be used in aqueous systems without any decrease in the activity. Overall, cholinium cation and amino acid anion, like Lys and Arg, containing BILs, are commonly used as they provide better yields due to their high number of hydrogen bonding sites that provide strong interaction with cellulose, in contrast to other ILs (Piedade et al. 2022).

4.4.2 BILs in Energy Storage

BILs behave differently structurally and dynamically in confined spaces, such as nanopores, than in bulk systems. These modifications significantly impact important characteristics like conductivity and viscosity, including creating ordered ionic layers and compact ion pairs. It is imperative to comprehend and optimize these effects to improve the performance of ILs in energy storage devices and make them more appropriate for sustainable applications. Because of this compatibility, BILs may be used as environmentally friendly electrolytes in MXene-based energy storage devices, including batteries and supercapacitors. The results demonstrate the possibility of combining cutting-edge 2D materials with biodegradable ionic liquids to create high-performing, environment-friendly energy storage systems (Moosavi et al. 2023).

Another study explores the interactions between two-dimensional titanium carbide (MXene) nanopores and biodegradable BILs. Using molecular dynamics simulations, the study showed that when BILs are confined inside MXene nanopores, they display unique structural and dynamical behaviours, with differences between neutral and charged environments. These interactions impact layering, mobility, and ion distribution—all crucial elements influencing supercapacitor performance. According to the results, the confinement effects of BILs can be maximized by adjusting the surface charge of MXene nanopores, improving the sustainability and efficiency of energy storage devices. There is high potential for combining cutting-edge nanomaterials with BILs to create high-performance, environmentally friendly supercapacitors (Nikpour et al. 2023).

4.4.3 BILs as Lubricants or Lubricant Additives

BILs can be used as base lubricants or as additives to traditional lubricants to improve performance. Their promising tribological properties, such as good oxidative and thermal stability and advantageous tribochemical interactions, have been proven beneficial to their lubricant activity. ILs have several benefits, including the following: their electrical conductivity lowers contact resistance in electrical systems; their tunable properties make them appropriate for a variety of tribosystems; their high chemical and thermal stability allows for use in harsh environments, and lastly, their low vapour pressure supports vacuum and safe applications because they are non-flammable and emit no emissions; and some ILs have high thermal conductivity, which helps dissipate heat during sliding process. Because of these characteristics, ILs are very appealing for cutting-edge and environmentally friendly lubrication technologies (Stolte et al. 2012).

The effectiveness of tributyl(methyl)phosphonium diphenylphosphate IL as an eco-friendly additive in biodegradable lubricants is examined in a study. When mixed into biodegradable base stocks, this BIL showed better antiwear qualities than traditional additives like amine phosphate. Tribological tests, such as reciprocating friction evaluations and pin-on-disc tests, demonstrated that BIL creates a strong tribofilm on steel surfaces, which was validated by methods like SEM–EDS and AFM. Furthermore, elastohydrodynamic tests showed that BIL is more capable of forming films than amine phosphate. This showed that tributyl(methyl)phosphonium diphenylphosphate is a BIL that shows promise for improving the sustainability and performance of biodegradable lubricants (Khemchandani et al. 2014).

4.5 Future Directions in Ionic Liquids

Several interesting research paths are arising in line with the structural, synthetic, and application-based developments described in this work as IL chemistry develops. Foundations of IL chemistry-including their structure, classification, synthesis, and physicochemical properties- future study will probably centre on designing task-specific ionic liquids (TSILs) with tunable functionality. Molecular engineering of these ILs allows one to customize solubility, viscosity, polarity, and thermal/electrochemical stability for particular uses. Understanding structure–property relations and dynamic behaviour in complex environments will depend critically on advanced characterization methods, including spectroscopy and high-resolution NMR. Recently, studies have been increasingly done with the help of computational design and AI-assisted discovery to invent new types of ionic liquids. Today, predicting the viscosity, conductivity, and solubility of ILs is done with machine learning and quantum chemical methods, cutting down on the need for trial-and-error in IL synthesis. There is increasing interest in ionic liquid crystals (ILCs) and deep eutectic solvents (DESs) as new alternatives, letting scientists control them like

ILs and offering improved biocompatibility and lower costs. There is also ongoing development of IL-based systems that can react to different conditions, which allows them to serve in sensing and adaptable materials.

The future of ILs in catalysis in green synthesis techniques is to incorporate them more in catalytic cycles and inert solvents. Because of their capacity to stabilize reactive intermediates and promote phase separation, ILs provide special media for catalytic, organic, photochemical, and biocatalytic reactions. Integrating enzyme-compatible IL systems to enable environmentally friendly biotransformations and creating recyclable IL-based catalysts for atom-efficient and waste-free transformations. Similarly, ILs in electrochemical and photochemical synthesis are expected to become more significant, especially for producing high-value chemicals and intermediates in mild, energy-efficient conditions.

ILs that increase reaction selectivity, decrease purification steps and lessen environmental impact will be the focus of future research in pharmaceuticals and fine chemical synthesis. ILs will be developed to solubilize and fractionate lignocellulosic biomass more efficiently in biomass conversion and biorefineries, facilitating bio-economically sustainable processes. For their non-volatility, ionic conductivity, and electrochemical stability, ILs will be essential in the energy sector's next-generation batteries, fuel cells, and supercapacitors, which is an excellent front. Additionally, ILs are being researched for their capacity to degrade persistent organic pollutants, remove heavy metals, and capture CO_2 in environmental remediation; future research can focus on making them economically feasible, recyclable, and have a low ecological impact.

IL-based nano-catalysis can stabilize active metal nanoparticles within IL matrices to create highly active and selective systems for organic transformations and energy conversion reactions. The synergistic effects between ILs and nanosurfaces can be explored using in situ analytical methods and computational modeling. Optimizing ILs as electrolytes in batteries, supercapacitors, and hydrogen storage systems with an emphasis on characteristics like broad electrochemical windows, thermal tolerance, and non-flammability will be a key focus for sustainable energy applications. Finally, the development of biodegradable ionic liquids will accelerate, emphasizing the creation of ILs from renewable feedstocks that maintain high performance while guaranteeing low environmental persistence.

In conclusion, the pressing need for safer, more effective, and environmentally friendly chemical processes is driving the highly interdisciplinary nature of IL research in the future. The convergence of sustainable engineering, materials science, and synthetic innovation in chemistry, biology, and the environment will guide the next generation of IL technologies.

References

An, X., Wang, P., Ma, X., Du, X., Hao, X., Yang, Z., & Guan, G. (2023). Application of ionic liquids in CO2 capture and electrochemical reduction: A review. *Carbon Resources Conversion*, *6*(2), 85–97. https://doi.org/10.1016/j.crcon.2023.02.003

Aziz, T., Alqarni, A. S., Ullah, H., Shehzad, F. K., Sohail, M., Tariq, M., Khan, K. A., & Asif, H. M. (2025). Synthesis of heterogeneous nanocatalysts comprising polyoxometalate ionic liquid-based magnetic nanocomposites for oxidation of thiobenzoic acid and 2,2′-dinitro-5,5′-dithiodibenzoic acid. *New Journal of Chemistry*, *49*(3), 990–1001. https://doi.org/10.1039/D4NJ03561E

Bakhshali-Dehkordi, R., Ghasemzadeh, M. A., & Safaei-Ghomi, J. (2020). Preparation and characterization of a novel DABCO-based ionic liquid supported on Fe 3 O 4 @TiO 2 nanoparticles and investigation of its catalytic activity in the synthesis of quinazolinones. *Applied Organometallic Chemistry*, *34*(9). https://doi.org/10.1002/aoc.5721

Bhowmik, D. D., Gupta, J. S., Kamandar, P. A., Gholap, A. D., Rojekar, S., & Hatvate, N. T. (2025). Dendrimers-Based Hydrogels and Nanogels for Drug Delivery. *Current Pharmaceutical Biotechnology*, *26*. https://doi.org/10.2174/0113892010347680250209183005

Bhowmik, D., Kamandar, P., Gupta, J., Mete, S., & Hatvate, N. (2024). An efficient, green and micellar catalyzed preparative-scale synthesis of 3,4-Disubstituted isoxazole-5(4H)-ones in the water. *Sustainable Chemistry for the Environment*, *5*, 100070. https://doi.org/10.1016/j.scenv.2024.100070

Bistgani, A. M., & Moradi, L. (2024). Immobilization of ionic liquids (ILs) on magnetic mesoporous silica nanotubes: A high throughput and reusable nanocatalyst for the green synthesis of indeno[1,2-b]indolone derivatives. *Microporous and Mesoporous Materials*, *372*, 113113. https://doi.org/10.1016/j.micromeso.2024.113113

Bohre, A., Modak, A., Chourasia, V., Ram Jadhao, P., Sharma, K., & Kishore Pant, K. (2022). Recent advances in supported ionic liquid catalysts for sustainable biomass valorization to high-value chemicals and fuels. *Chemical Engineering Journal*, *450*, 138032. https://doi.org/10.1016/j.cej.2022.138032

Bolarinwa Solanke, Femi Bamidele Onita, Obinna Joshua Ochulor, & Henry Oziegbe Iriogbe. (2024). Advances in CO2 injection and monitoring technologies for improved safety and efficiency in CCS projects. *International Journal of Frontline Research in Engineering and Technology*, *2*(1), 031–040. https://doi.org/10.56355/ijfret.2024.2.1.0020

Chen, Y., & Mohsen Sadeghzadeh, S. (2023). Dendritic fibrous nano-titanium (DFNT) with highly dispersed poly(ionic liquids) as a nanocatalyst for synthesis of dimethyl carbonate from methanol and carbon dioxide. *Journal of Molecular Liquids*, *384*, 122201. https://doi.org/10.1016/j.molliq.2023.122201

Choudhary, G., Dhariwal, J., Saha, M., Trivedi, S., Banjare, M. K., Kanaoujiya, R., & Behera, K. (2023). Ionic liquids: environmentally sustainable materials for energy conversion and storage applications. *Environmental Science and Pollution Research*, *31*(7), 10296–10316. https://doi.org/10.1007/s11356-023-25468-w

Dr. D. Valli Sowbhagyam. (2024). Ionic Liquids as Green Solvents: A Comprehensive Review. *International Research Journal on Advanced Engineering Hub (IRJAEH)*, *2*(02), 220–224. https://doi.org/10.47392/IRJAEH.2024.0035

Faraji, S., Shekaari, H., Zafarani-Moattar, M. T., Mokhtarpour, M., & Asghari, E. (2023). Thermal properties of phase change materials ionic liquid/fatty acids for thermal energy storage applications. *Journal of Energy Storage*, *67*, 107464. https://doi.org/10.1016/j.est.2023.107464

Gadkari, Yatin U., Hatvate, N. T., & Telvekar, V. N. (2021a). Concentrated solar radiation-assisted one-pot/multicomponent synthesis of pyranopyrazole derivatives under neat condition. *Research on Chemical Intermediates*, *47*(10), 4245–4255. https://doi.org/10.1007/s11164-021-04530-7

Gadkari, Yatin U., Hatvate, N. T., & Telvekar, V. N. (2021b). Solar energy as a renewable energy source for preparative-scale as well as solvent and catalyst-free Hantzsch reaction. *Sustainable Chemistry and Pharmacy*, *21*, 100444. https://doi.org/10.1016/j.scp.2021.100444

Gadkari, Y.U., Jadhav, N. L., Hatvate, N. T., & Telvekar, V. N. (2020). Concentrated Solar Radiation Aided Green Approach for Preparative Scale and Solvent-Free Synthesis of 3-Methyl-4-(hetero)arylmethylene Isoxazole-5(4H)-ones. *ChemistrySelect*, *5*(39), 12320–12323. https://doi.org/10.1002/slct.202003348

Gaete, J., Valdebenito, G., Moglia, I., Morales-Verdejo, C., Aguirre, P., Fernandes, J. A., & Abarca, G. (2025). Copper-based magnetic nanocatalysts for the catalytic transfer hydrogenation of biomass-derived furfural. *Applied Surface Science*, *689*, 162566. https://doi.org/10.1016/j.apsusc.2025.162566

Gaete, J., Valdebenito, G., Valdebenito, C., Morales-Verdejo, C., Wrighton-Araneda, K., Aguirre, P., & Abarca, G. (2024). Tuning C-N Ullmann coupling with anion-exchanged ionic liquid-functionalized nanoparticle catalysts. *Journal of Molecular Liquids*, *403*, 124782. https://doi.org/10.1016/j.molliq.2024.124782

Gao, Y.-R., Cao, J.-F., Shu, Y., & Wang, J.-H. (2021). Research progress of ionic liquids-based gels in energy storage, sensors and antibacterial. *Green Chemical Engineering*, *2*(4), 368–383. https://doi.org/10.1016/j.gce.2021.07.012

Georgijević, J., Milikić, J., Aykut, Y., Zdolšek, N., Santos, D. M. F., Bayrakçeken, A., & Šljukić, B. (2024). Fe, Cu-decorated carbon material produced from ionic liquids as resourceful electrocatalyst for water splitting. *Journal of Electroanalytical Chemistry*, *967*, 118455. https://doi.org/10.1016/j.jelechem.2024.118455

Gholap, A. D., Gupta, J., Kamandar, P., Bhowmik, D. D., Rojekar, S., Faiyazuddin, Md., Hatvate, N. T., Mohanto, S., Ahmed, M. G., Subramaniyan, V., & Kumarasamy, V. (2024a). Harnessing Nanovaccines for Effective Immunization–A Special Concern on COVID-19: Facts, Fidelity, and Future Prospective. *ACS Biomaterials Science & Engineering*, *10*(1), 271–297. https://doi.org/10.1021/acsbiomaterials.3c01247

Gholap, A. D., Kapare, H. S., Pagar, S., Kamandar, P., Bhowmik, D., Vishwakarma, N., Raikwar, S., Garkal, A., Mehta, T. A., Rojekar, S., Hatvate, N., & Mohanto, S. (2024b). Exploring modified chitosan-based gene delivery technologies for therapeutic advancements. *International Journal of Biological Macromolecules*, *260*, 129581. https://doi.org/10.1016/j.ijbiomac.2024.129581

Gholap, A. D., Pardeshi, S. R., Hatvate, N. T., Dhorkule, N., Sayyad, S. F., Faiyazuddin, M., & Khalid, M. (2024c). Environmental implications and nanotechnological advances in octocrylene-enriched sunscreen formulations: A comprehensive review. *Chemosphere*, *358*, 142235. https://doi.org/10.1016/j.chemosphere.2024.142235

Gholap, A. D., Khuspe, P. R., Faiyazuddin, M., Uddin, M. J., Gupta, J., Kamandar, P., Bhowmik, D. D., & Hatvate, N. T. (2025). Current scenario and challenges of lipid-based nanomedicines for pulmonary drug delivery. In Lipids in Pulmonary Drug Delivery (pp. 21–47). Elsevier. https://doi.org/10.1016/B978-0-443-22374-7.00002-5

Gholinejad, M., Iranpanah, M., Karimi, S., & Sansano, J. M. (2024). Cysteine and ionic liquid modified magnetic nanoparticles supported RuCu as a new bimetallic catalyst in reduction reactions. *Journal of Molecular Structure*, *1298*, 137100. https://doi.org/10.1016/j.molstruc.2023.137100

Hamidinasab, M., Bodaghifard, M. A., & Mobinikhaledi, A. (2020). Green synthesis of 1 H - pyrazolo[1,2- b]phthalazine-2-carbonitrile derivatives using a new bifunctional base–ionic liquid hybrid magnetic nanocatalyst. *Applied Organometallic Chemistry*, *34*(2). https://doi.org/10.1002/aoc.5386

Hatvate, N., & Gupta, J. (2024). An efficient, green and solvent-free three-component synthesis of 2,4,6-triarylpyridines utilizing ZSM-5 as a heterogeneous catalyst. *Synthetic Communications*, *54*(20), 1760–1770. https://doi.org/10.1080/00397911.2024.2403779

Hatvate, Navnath T., & Ghodse, S. M. (2020). One-pot three-component synthesis of isoxazole using ZSM-5 as a heterogeneous catalyst. *Synthetic Communications*, *50*(23), 3676–3683. https://doi.org/10.1080/00397911.2020.1815786

Hatvate, Navnath T., Ghodse, S. M., Mundlod, K. N., & Telvekar, V. N. (2020). Metal-Free Synthesis of Pyrimidinone Derivatives via Biginelli Reaction Using Aqueous NaICl2. *Letters in Organic Chemistry, 17*(8), 613–617. https://doi.org/10.2174/1570178617666191126095808

Hatvate, Navnath T., Ghodse, S. M., & Telvekar, V. N. (2019). ZSM-5 catalyzed one pot three-component synthesis of 5-substituted-1H-tetrazoles from aldehyde. *Synthetic Communications, 49*(24), 3553–3559. https://doi.org/10.1080/00397911.2019.1679185

Hatvate, N.T., Ghodse, S. M., & Telvekar, V. N. (2018). Metal-free synthesis of 2-aminothiadiazoles via TBHP-Mediated oxidative C-S bond formation. *Synthetic Communications, 48*(3), 285–290. https://doi.org/10.1080/00397911.2017.1398330

Islam, M. R., Khan, M. A., Ali, M., & Sk, M. P. (2023). Progressions in ionic liquid-based electrochemical research. In Advanced Applications of Ionic Liquids (pp. 3–21). Elsevier. https://doi.org/10.1016/B978-0-323-99921-2.00019-7

Jaddan, R. I., Kadium, A. H., Mahmood, A. A., Shather, A. H., Al-Bayati, A. D. J., Shaghnab, M. L., & Almashhadani, H. A. (2023). Effectual and rapid synthesis of bis(pyrazolyl)methanes and pyrazolopyranopyrimidines catalyzed by nano-[TSPSED][Cl]2 and investigating their biological activity. *Research on Chemical Intermediates, 49*(9), 4103–4126. https://doi.org/10.1007/s11164-023-05060-0

Jia, L., Yu, H., Liu, J., Zhu, B., Zhang, S., Tian, J., Xie, H., & Zheng, W. (2024). Ionic liquid-assisted controlled synthesis of multi-site Ni2P as a bifunctional catalyst for electrocatalytic water splitting. *International Journal of Hydrogen Energy, 88*, 1267–1279. https://doi.org/10.1016/j.ijhydene.2024.09.120

Kadhim, M. M., Tabarsaei, N., Ghorchibeigi, M., & Sadeghi Meresht, A. (2023). New MCRs in Ionic Liquid: Green Synthesis and Biological Activity Investigation of New Pyrazoloazepines: Application of Ag/Fe3O4/CdO@MWCNT MNCs in Reduction of Organic Pollutant. *Polycyclic Aromatic Compounds, 43*(7), 5785–5806. https://doi.org/10.1080/10406638.2022.2106253

Karuppasamy, K., Theerthagiri, J., Vikraman, D., Yim, C.-J., Hussain, S., Sharma, R., Maiyalagan, T., Qin, J., & Kim, H.-S. (2020a). Ionic Liquid-Based Electrolytes for Energy Storage Devices: A Brief Review on Their Limits and Applications. *Polymers, 12*(4), 918. https://doi.org/10.3390/polym12040918

Karuppasamy, K., Vikraman, D., Hwang, I.-T., Kim, H.-J., Nichelson, A., Bose, R., & Kim, H.-S. (2020b). Nonaqueous liquid electrolytes based on novel 1-ethyl-3-methylimidazolium bis (nonafluorobutane-1-sulfonyl imidate) ionic liquid for energy storage devices. *Journal of Materials Research and Technology, 9*(2), 1251–1260. https://doi.org/10.1016/j.jmrt.2019.11.052

Kaur, G., Sharma, S., Singh, M. D., Nalwa, K. S., Sivasubramanian, S. C., & Dalvi, A. (2024). Ionic liquid composites with garnet-type Li6.75Al0.25La3Zr2O12: Stability, electrical transport, and potential for energy storage applications. *Materials Chemistry and Physics, 317*, 129205. https://doi.org/10.1016/j.matchemphys.2024.129205

Khemchandani, B., Somers, A., Howlett, P., Jaiswal, A. K., Sayanna, E., & Forsyth, M. (2014). A biocompatible ionic liquid as an antiwear additive for biodegradable lubricants. *Tribology International, 77*, 171–177. https://doi.org/10.1016/j.triboint.2014.04.016

Krishnan, D., Schill, L., Axet, M. R., Philippot, K., & Riisager, A. (2024). Nanocatalysts Based on Ni Nanoparticles in Functionalized Ionic Liquids for Chemoselective Hydrogenation and Transfer Hydrogenation of α,β-unsaturated Carbonyl Compounds. *ChemCatChem, 16*(4). https://doi.org/10.1002/cctc.202301441

Li, H., Zhu, X., He, T., Qu, C., Tian, Q., Xie, H., Hu, L., Liang, S., Zhang, L., & Yuan, J. (2024). Hydrophilic–Hydrophobic Regulation of the Cellulose Suspending Ionic Liquid Hydrogel/Palladium Nanocomposite Catalyst for Hydrogenation. *ACS Sustainable Chemistry & Engineering, 12*(14), 5390–5401. https://doi.org/10.1021/acssuschemeng.3c03168

Li, S., Lindsay, H., Mannari, V., & Texter, J. (2023). Liquid polymerized ionic liquids for energy storage applications. *Journal of Molecular Liquids, 376*, 121403. https://doi.org/10.1016/j.molliq.2023.121403

Liu, Y., Li, Y., Wu, H., Xu, S., Zhang, B., Li, S., Du, R., Jiang, M., Chen, Z., Lv, Y., & Wang, Z.-G. (2024). Robust Oxidase-Mimetic Supramolecular Nanocatalyst for Lignin Biodegradation. *Nano Letters, 24*(8), 2520–2528. https://doi.org/10.1021/acs.nanolett.3c04505

Lv, X., Lv, A., Xie, T., Shao, Z., Yin, G., Li, D., Xu, L., & Sun, S. (2024). Enhanced Stability and Catalytic Activity of a Nanocatalyst with Reusable Ionic Liquid Hydrogels for the Reduction of Organic Pollutants. *Langmuir, 40*(4), 2210–2219. https://doi.org/10.1021/acs.langmuir.3c03275

MacFarlane, D. R., Tachikawa, N., Forsyth, M., Pringle, J. M., Howlett, P. C., Elliott, G. D., Davis, J. H., Watanabe, M., Simon, P., & Angell, C. A. (2014). Energy applications of ionic liquids. *Energy Environ. Sci., 7*(1), 232–250. https://doi.org/10.1039/C3EE42099J

Mim, M., Habib, K., Farabi, S. N., Zaed, M. A., & Saidur, R. (2025). Investigation of the thermophysical properties of PCMs with novel ionic liquid assisted nanocomposite for sustainable thermal energy storage application. *Case Studies in Thermal Engineering, 70*, 106117. https://doi.org/10.1016/j.csite.2025.106117

Mirzaei-Saatlo, M., Asghari, E., Shekaari, H., Pollet, BG., & Vinodh, R. (2024). Performance of ethanolamine-based ionic liquids as novel green electrolytes for the electrochemical energy storage applications. *Electrochimica Acta, 474*, 143499. https://doi.org/10.1016/j.electacta.2023.143499

Monavari, N., Zhiani, R., Hosseiny, M., Khosroyar, S., Ebrahimi, Z., & Moradi, M. (2024). Fibrous phosphosilicate with highly dispersed poly(ionic liquids) as a nanocatalyst for production of biopolymer from limonene epoxide and CO2. *Journal of CO2 Utilization, 90*, 102978. https://doi.org/10.1016/j.jcou.2024.102978

Moosavi, M., Torkzadeh, M., & Nikpour, A. (2023). Physisorption of biodegradable choline-based ionic liquids (CBILs) and their aqueous solutions on 2D titanium carbide (MXene) nanosheets as promising media in energy storage systems. *Journal of Molecular Liquids, 382*, 121768. https://doi.org/10.1016/j.molliq.2023.121768

Moradi, L., Sasi, H. R., & Dehghani, A. (2024). Introducing a high throughput nanocatalytic method toward the synthesis of some pyrazolo phthalazine derivatives under green conditions utilizing imidazoliume based ionic liquid supported on the silica-coated nanosized perlite as a novel, reusable and eco. *Research on Chemical Intermediates, 50*(4), 1619–1643. https://doi.org/10.1007/s11164-024-05234-4

Mousapour, M., & Shirini, F. (2024). Comparison of the effectiveness of two piperazine based nanocatalysts in the synthesis of benzoxazoles and benzimidazoles and use of the powerful one in the N -Boc protection of amines. *RSC Advances, 14*(13), 9218–9227. https://doi.org/10.1039/D4RA01300J

Nasri, A., Jaleh, B., Shabanlou, E., Nasrollahzadeh, M., Ali Khonakdar, H., & Kruppke, B. (2022). Ionic liquid-based (nano)catalysts for hydrogen generation and storage. *Journal of Molecular Liquids, 365*, 120142. https://doi.org/10.1016/j.molliq.2022.120142

Nikpour, A., Moosavi, M., & Torkzadeh, M. (2023). Structural, spectral, and dynamical behaviors of biodegradable choline-based ionic liquids (CBILs) confined inside neutral and charged mxene nanopores for supercapacitor applications. *Colloids and Surfaces A: Physicochemical and Engineering Aspects, 674*, 131968. https://doi.org/10.1016/j.colsurfa.2023.131968

Pardeshi, S. R., Gholap, A. D., Hatvate, N. T., Gharat, K. D., Naik, J. B., & Omri, A. (2024). Advances in dorzolamide hydrochloride delivery: harnessing nanotechnology for enhanced ocular drug delivery in glaucoma management. *Discover Nano, 19*(1), 199. https://doi.org/10.1186/s11671-024-04154-x

Piedade, P. J., Kochańska, E., & Lukasik, R. M. (2022). Biodegradable ionic liquids in service of biomass upgrade. *Current Opinion in Green and Sustainable Chemistry, 35*, 100609. https://doi.org/10.1016/j.cogsc.2022.100609

Ray, A., & Saruhan, B. (2021). Application of Ionic Liquids for Batteries and Supercapacitors. *Materials, 14*(11), 2942. https://doi.org/10.3390/ma14112942

Sangeetha, B., Pravin, R., & Baskar, G. (2024). Technoeconomic and environmental impact assessment of industrial scale biodiesel production process from algal oil using a reusable basic ionic liquid magnetic nanocatalyst. *Energy Conversion and Management, 314*, 118668. https://doi.org/10.1016/j.enconman.2024.118668

Šibalić, D., Šalić, A., Zelić, B., Tran, N. N., Hessel, V., Nigam, K. D. P., & Tišma, M. (2023). Synergism of ionic liquids and lipases for lignocellulosic biomass valorization. *Chemical Engineering Journal, 461*, 142011. https://doi.org/10.1016/j.cej.2023.142011

Singh, A., Dhapola, P. S., Kumar, S., Konwar, S., Singh, P. K., Ali, A. M. M., Agarwal, D., Singh, D., Strzałkowski, K., Yahya, M. Z. A., Savilov, S. V., & Joshi, B. (2022a). Highly conducting ionic liquid doped polymer electrolyte for energy storage applications. *Journal of Science: Advanced Materials and Devices, 7*(4), 100511. https://doi.org/10.1016/j.jsamd.2022.100511

Singh, V., Amirchand, K. D., & Gardas, R. L. (2022b). Ionic liquid-nanoparticle based hybrid systems for energy conversion and energy storage applications. *Journal of the Taiwan Institute of Chemical Engineers, 133*, 104237. https://doi.org/10.1016/j.jtice.2022.104237

Stolte, S., Steudte, S., Areitioaurtena, O., Pagano, F., Thöming, J., Stepnowski, P., & Igartua, A. (2012). Ionic liquids as lubricants or lubrication additives: An ecotoxicity and biodegradability assessment. *Chemosphere, 89*(9), 1135–1141. https://doi.org/10.1016/j.chemosphere.2012.05.102

Telvekar, V. N., Mundlod, K. N., Jadhav, G. B., Hatvate, N. T., & Ghodse, S. M. (2020). Synthesis of Novel molecular hybrid 3-(4,5-dihydro-1H-pyrazol-3-yl)-4-hydroxy-2H-chromen-2-one derivatives and In Vitro α-glucosidase and α-amylase inhibitory activity. *Research Journal of Pharmacy and Technology, 13*(10), 4529. https://doi.org/10.5958/0974-360X.2020.00798.2

Wolny, A., & Chrobok, A. (2021). Ionic Liquids for Development of Heterogeneous Catalysts Based on Nanomaterials for Biocatalysis. *Nanomaterials, 11*(8), 2030. https://doi.org/10.3390/nano11082030

Zenner, J., Kang, L., Antil, N., Johny, J., DeBeer, S., Leitner, W., & Bordet, A. (2025). Bimetallic Mn x Ru 100– x Nanoparticles on Supported Ionic Liquid Phases (Mn x Ru 100– x @SILP) as Tunable Hydrogenation Catalysts. *ACS Catalysis, 15*(4), 3227–3235. https://doi.org/10.1021/acscatal.4c05494

Zhao, W., Fan, C., Liu, L., Ma, X., & Feng, B. (2024). A novel non-aqueous absorbent of amino-functionalized ionic liquid for efficient CO2 capture. *Chemical Engineering Research and Design, 209*, 102–110. https://doi.org/10.1016/j.cherd.2024.07.057

Zhou, T., Gui, C., Sun, L., Hu, Y., Lyu, H., Wang, Z., Song, Z., & Yu, G. (2023). Energy Applications of Ionic Liquids: Recent Developments and Future Prospects. *Chemical Reviews, 123*(21), 12170–12253. https://doi.org/10.1021/acs.chemrev.3c00391

MIX
Papier aus verantwortungsvollen Quellen
Paper from responsible sources
FSC® C105338

If you have any concerns about our products,
you can contact us on
ProductSafety@springernature.com

In case Publisher is established outside the EU,
the EU authorized representative is:
Springer Nature Customer Service Center GmbH
Europaplatz 3, 69115 Heidelberg, Germany

Printed by Libri Plureos GmbH
in Hamburg, Germany